# Patrick Moore's
# Practical Astronomy Series

I0911190

## Springer
*London*
*Berlin*
*Heidelberg*
*New York*
*Barcelona*
*Hong Kong*
*Milan*
*Paris*
*Singapore*
*Tokyo*

# Other titles in this series

# Astronomy with Small Telescopes

## Up to 5-inch, 125mm

Stephen F. Tonkin (Ed.)

With 75 Figures

Springer

Stephen F. Tonkin, BSc, FRAS
66 Earlswood Drive, Alderholt, Fordingbridge,
Hampshire SP6 3EN, UK

*Photographs:* All photographs shown in the text are by the
contributors.

Patrick Moore's Practical Astronomy Series ISSN 1431-9756

ISBN 1-85233-629-3 Springer-Verlag London Berlin Heidelberg

British Library Cataloguing in Publication Data
Astronomy with small telescopes: up to 5-inch, 125 mm. –
    (Patrick Moore's practical astronomy series)
    1. Telescopes
    I. Tonkin, Stephen F., 1950 – II. Moore, Patrick, 1923–
    522.2
ISBN 1852336293

Library of Congress Cataloging-in-Publication Data
Astronomy with small telescopes up to 5 inch, 125mm /
    Stephen F. Tonkin, ed.
        p.    cm. – (Patrick Moore's practical astronomy series,
            ISSN 1431–9756)
    ISBN 1–85233–629–3 (alk. paper)
    1. Telescopes. I. Tonkin, Stephen F., 1950– II. Series.
QB88.A779 2001
522'.2–dc21                                          00–049220

© Springer-Verlag London Limited 2001
Printed in Great Britain

Typeset by EXPO Holdings, Malaysia
Printed at the Cromwell Press Ltd., Trowbridge, Wiltshire
58/3830-543210  Printed on acid-free paper  SPIN 10732269

This book is for my father, Allan Tonkin, who, by his example, taught me the distinction between price and value.

# Contents

# III  **Catadioptrics**

# IV  **Radio**

# Introduction

Part of the history of telescopic astronomy has been a plea that is embodied in the famous last words of Goethe: "More light!" Only 50 years ago, a 150mm (6-inch) reflector or a 75mm (3-inch) refractor was considered to be a good-sized amateur instrument, and anything significantly larger was restricted to either the wealthy or the skilled amateur optician.

During the last quarter-century, innovations in materials, design and manufacturing have delivered increasingly larger instruments into the hands of amateurs. Perhaps the most influential of these has been the genius for simplicity that led John Dobson to develop and champion his eponymous telescope mounting. Other amateurs built on Dobson's ideas, adding equatorial tracking, computer control, and lightweight or collapsible design. Mass production, in particular in the Far East, has driven down the prices of conventional telescopes and of many other items of astronomical equipment. The plea for more light had been answered and, as we enter the 21st century, amateur telescopes with four times the light-gathering capacity of those of only a few decades ago are commonplace. It is no longer unusual for amateurs to own reflectors with an aperture of half a metre (20 inches) or more, and the advice often given to those seeking advice on a first telescope is to buy a 200mm (8-inch) reflector.

There is a downside to this move to larger apertures. Merely by virtue of their size, large instruments are less portable than an equivalent smaller instrument, and they tend to be more expensive.[1] As the quality of the night skies around centres of population has become increasingly degraded, the need for

---

[1] This last factor does not, however, always follow. When Celestron introduced its 125mm (5-inch) Schmidt–Cassegrain telescope, the C5, in 1971, the production cost was higher than that of its 200mm (8-inch) big brother!

portability has grown stronger. Increasing ease of world travel leads amateurs to "chase" localised phenomena like solar eclipses. It also makes it easier for northern-hemisphere amateurs to head south and sample the many delights of the southern-hemisphere skies (and vice versa). Consequently much effort has been put into making large amateur telescopes more portable. Large ultra-portable telescopes are usually custom-built, and more often than not are made by an enthusiast who has the necessary design and workshop skills. Similar commercial models, when they exist, tend to be considerably more expensive than "standard" instruments.

Recently another trend has been emerging. Anyone who peruses the advertisements in astronomical periodicals will have noticed that since the mid-1990s there has been an increase in the range of small telescopes available to the amateur. There is a wide range in the quality of these instruments, and this is reflected in the price. However, the upshot is that portable instruments of reasonable quality are now available at a price that most amateur astronomers can afford.

Small instruments have always occupied a specialist niche – the 90mm Questar is perhaps the small telescope with the best reputation for optical and mechanical quality – but the recent "downsizing revolution" has led to small telescopes being the telescope of choice for some serious observers. Price is an important factor, but portability and convenience are even more important. When I want to escape the light pollution of the village in which I live, my entire portable set-up, complete with imaging equipment and a host of accessories, fits neatly into the boot (trunk) of our small family car with room to spare. If I wish to observe off the beaten track, the telescope and driven tabletop mount fits into my backpack, again with room to spare. On a recent eclipse trip with my family, where space in the car was at a premium, the small refractor that is my ultra-portable telescope squeezed in almost unnoticed along one side of the roof-box.

The portability is obvious, but does it impose limitations on the capability of the telescope? Of course a small telescope has its limitations – even remarkable quality cannot overcome the laws of optical physics – but they are not so severe as to prevent a lifetime of useful, pleasurable amateur observing. Remember that, fifty years ago, many of today's small telescopes would have been considered to be improvements on those that were doing useful work at the small-aperture end of the serious amateur telescope range.

An alternative to increasing aperture in order to gather more light is to increase the efficiency with which light is gathered. Modern lens coatings do this to some extent, but the greatest advance in this area has been in the realm of photography, now perhaps better called "astro-imaging". A modest amateur CCD camera gathers light much more efficiently than does conventional photographic emulsion, and, where light-gathering is concerned, a CCD camera on a 100mm (4-inch) telescope will "grab photons" at approximately the same rate as photographic emulsion on an equivalent telescope with ten times the aperture. CCDs do not yet have the resolution of photographic film, and a 100mm does not have the resolution of a 1-metre (assuming that the seeing allows the larger telescope to be used to its full capability), but the photographic potential of a small amateur telescope is far greater than it was twenty years ago.

If it is to be useful for visual work, a small instrument *demands* that you hone and optimise your observing techniques. In the first section of this book, Jay Freeman shows how the small telescope can be used in that realm – observation of deep sky objects – which is so often mistakenly believed to be the sole preserve of large telescopes. The skills you will learn are directly transferable to a larger telescope if you use one.

Jay Freeman uses sturdily mounted telescopes of very high optical quality. Many of the less expensive small telescopes available today do not necessarily share the same degree of excellence; it would be unreasonable to expect it at the price at which they are sold. Dave Mitsky shows what can be done with an inexpensive refractor, whilst Kevin Daly and Dwight Elvey examine the possibilities of small reflectors.

Inexpensive small telescopes are those that are most likely to come into the hands of youngsters who embark on this avocation that we share. Tim Tonkin is a young person who is deriving great pleasure from using his small reflector as he develops his skills and knowledge. His telescope combines all the benefits of a small telescope with the simplicity of a Dobsonian. The other telescope that is most likely to come into the hands of beginners, especially youngsters, is the ubiquitous (and frequently denigrated) 60mm refractor. Rob Hatch and I both have better examples of the genre and show how these can be used for serious astronomy. I also show how the less good examples can be improved and made, quite simply, into useful astronomical instruments.

No book on small telescopes would be complete without mention of the small catadioptrics that are taking an increasing market share. Mike Weasner calls his little Maksutov–Cassegrain "The Mighty ETX" with good reason. The Maksutov design is especially suited to small telescopes, as the Questar showed some decades ago. Combining the design with modern materials has brought down the cost, and the 90mm ETX has found favour with both experienced observers, many of whom regard it as the ideal second/ultraportable telescope, and beginners, many of whom appreciate the computerisation that enables them to spend more time observing than hunting.

The 125mm Schmidt–Cassegrain has an excellent pedigree. During the last three decades it has come on a variety of mounts. This indicates its inherent flexibility and, the world over, the optical tube assembly is found in a variety of amateur applications, from simple visual observations mounted on a photo tripod, to imaging on an ultra-stable permanent mount. My "C5++" demonstrates how the basic telescope is amenable to "accessorisation", resulting in an extremely flexible combination that is applicable to a variety of applications.

Lastly, I show how astronomical radio observations can be made using readily available equipment.

With such a variety of small telescopes now available, and more being announced every few months, it is not possible to include a chapter on every single one of them. However, this book does address each *class* of small telescope and the use to which it can be put. But how does one define "small" in this context? For the purposes of this book I have used the arbitrary definition that small refractors and Maksutovs are those with apertures of less than 100 mm (4 inches), while small reflectors and Schmidt–Cassegrains are those with apertures of less than 150 mm (6 inches). Maksutovs and Schmidt–Cassegrains are distinguished in this way because the curvature of the Maksutov's meniscus lens makes this design less easily enlarged than the Schmidt.

This book is not intended as an instruction manual, but rather to give you some idea of what is attainable, and hopefully to inspire you to get the best out of your small telescope. Whether you do it for pleasure or for "serious" observation, I hope that you will join the growing band of users of small telescopes and share in our satisfaction with these underrated instruments.

Stephen Tonkin
Alderholt, January 2001

# Part I

# **Refractors**

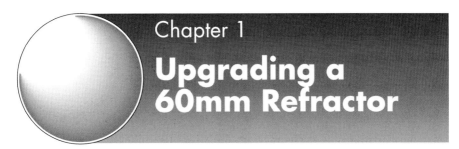

# Chapter 1
# Upgrading a 60mm Refractor

Stephen Tonkin

So, you have been told, or have discovered, that the quality of your new acquisition falls more than a little short of what the advertising and the graphics on the box led you to believe. Don't despair – most inexpensive 60mm (2.4-inch) refractors can be rendered much more useful with the expenditure of some time and a little money.

## What Can You Expect?

Firstly, you already have a very portable instrument which is easy to set up. Don't underestimate the value of this: it means that the telescope is likely to be used. There are many telescopes of better optical and mechanical quality which languish in sheds and garages, almost entirely unused, because they are heavy and are time-consuming to set up – by definition, any telescope which is not used, whatever the reason, is effectively useless.

No telescope can perform better than the laws of physics allow. A good 60mm telescope will enable you to observe point objects down to about 12th magnitude (with averted vision from a dark site), to achieve a resolution of 2 or 3 arc seconds (which is often better than atmospheric seeing will permit), and to sustain a magnification of around ×100 before the image starts falling apart (seeing permitting). In practice this means that you should be able to discern surface detail on Mars and Jupiter, distinguish the Cassini Division in the rings of Saturn, see sufficient detail on the Moon to make it exciting, observe sunspots (using a suitable objective filter or projection), watch and time lunar occultations (see Chapter 3), find detail in lots of open clusters, split many double stars, notice that globular clusters are at least globular, spot many small bright galaxies, pick out several planetary nebulae, and observe most (if not all) of the Messier objects and a heck of a lot else besides. That's quite a lot to be getting on with, even if it is only a fraction of what you can see with a bigger, better amateur instrument, especially where the faint fuzzies are concerned. You will *not* see volcanoes on Io, storms on Neptune, spiral arms in M101, colour in deep-sky objects, or anything like the pretty pictures on the package your telescope came in – but then, most of that stuff isn't visible in any amateur telescope!

All the observing tricks you learn with your 60mm will be transferable to your next, better, scope, *if* you can make using this one both fruitful and pleasurable. So, let's look at how you can push this telescope to its limit. What follows comes under the heading of "generic advice"; some of it may be inapplicable to your particular telescope and some of it may need adapting. It is not worth spending a great deal of money on a 60mm – you could easily spend a few hundred pounds (or dollars) upgrading it, and the money is probably better saved and put towards your next telescope.

# Optical Upgrades

## Eyepieces

Most of these telescopes are supplied with appalling eyepieces. Generally the only one worth keeping is the lowest-power one. The high-power ones are usually ghastly, with focal lengths which are far too short for these telescopes, except for the shorter focal ratio variants. I think it is worth getting a good-quality medium- or high(ish)-power eyepiece – this is the only money I'd be prepared to spend on optics for this type of telescope. For the older $f/15$ variety of 60mm refractor my choice for a high-power eyepiece is a 9mm Orthoscopic (×100), which is readily available and inexpensive. For the newer $f/11.2$ variety, an eyepiece with a focal length of around 7 mm is as short as I would consider. Note, however, that a high-power eyepiece will be frustrating to use unless the telescope mount is sufficiently steady.

If you are considering a medium-power eyepiece, use one that gives a magnification equal to the telescope aperture in mm, i.e. an eyepiece with a focal length equal to the focal ratio in mm. Thus, for a $f/11$ instrument get an eyepiece of around 11 mm focal length. A source of low-power eyepieces is broken binoculars from junk shops or car-boot sales (garage sales). They can be adapted to fit into holders made from the tubing from a Barlow lens or image erector (see Fig. 1.1).

Some people advise obtaining a hybrid star diagonal with these instruments, so that 1.25-inch (31.8mm) eyepieces can be used – the diagonal fits a 0.965-inch (24.5mm) focuser and takes 1.25-inch eyepieces. This sounds a good idea on the face of it, but there are a couple of potential pitfalls:

1. Hybrid diagonals tend to have longer optical paths than do the 0.965-inch ones. Similarly, the focal plane of 1.25-inch eyepieces can be farther back along the eyepiece barrel than in 0.965-inch eyepieces. As a consequence it can be impossible to bring the eyepiece to focus unless other modifications, such as shortening the tube, are made to the telescope. If you are considering the hybrid diagonal option, you would be advised to try it out on your telescope before committing yourself to any purchase.

2. One rationale given for the use of 1.25-inch eyepieces is that they generally have wider fields of view.

**Figure 1.1.**
Eyepieces. Left to right: the 20mm Huygenian supplied with the telescope, a good-quality 9mm orthoscopic, and a binocular eyepiece housed in part of the supplied Barlow fitting.

This can be negated by the 0.965-inch drawtube. Again, it is advisable to try before you buy.

If the "hybrid" path is a workable option, it may be worth considering if you intend to use the eyepieces you are collecting with a future telescope. But do remember that an eyepiece which gives good images at $f/15$ or $f/11$ may not do so at shorter focal ratios, where the more obtuse light-cone is far more demanding of eyepiece quality. If you are intending to acquire, say, an $f/6$ Newtonian, do check that the eyepiece is usable at this focal ratio as well.

## Barlow Lens

A Barlow lens is a diverging or negative lens which increases the effective focal length of an objective lens, thereby increasing the magnification. The idea is that two eyepieces and a Barlow will give you the same flexibility of magnification as will four eyepieces, and will give higher magnifications with less powerful eyepieces. By increasing the focal ratio, the Barlow lens reduces the angle of the light cone entering the eyepiece. The light cone therefore impinges upon the field lens of the eyepiece at a much smaller angle and over a smaller area; it is therefore much less demanding of eyepiece quality.

The idea is excellent, as long as the Barlow is also excellent, which the one supplied with your telescope almost certainly won't be. These are often singlet biconcave lenses which introduce chromatic aberration into the optical system and also give an enormously restricted field of view. The little biconcave lens itself, which can often be pushed out with a pencil, may be of interest to an optical tinkerer. More importantly, the tube can then be cut down and used to house homemade eyepieces (see above) or extension tubes (see Chapter 3).

## Sun Filter

Many of these telescopes are supplied with a "Sun filter" which screws into an eyepiece. On the face of it such filters ought to work, but in reality they are potentially very dangerous. Situated as they are close to the focus of the objective lens, the heat of the Sun is concentrated upon them. They work by absorbing the Sun's heat to prevent it from reaching the eye. But the filter can be rapidly heated to very high temperatures, causing it to crack, thus allowing the heat of the Sun through to the eyepiece and so to your eye.

There are only two things to do with this abomination. The first and simplest is to smash it with a hammer. However, you can get some educational value from it by waiting for a hot sunny day and "testing" it to destruction in the presence of some young observers. Instruct them that heads are to go nowhere near small end of the telescope, remove the threaded barrel from the Barlow or an eyepiece, screw the filter into it, place it in the focuser so that the Sun is focused onto it, cap the finder, track the Sun, and wait. Everyone will learn a valuable lesson which may, in the future, save someone's eyesight.

If you wish to observe the Sun, by far the safest way is to project its image onto a screen. The Huygenian eyepiece which is usually provided with this sort of telescope is ideal for projection, since it contains no cemented elements. (The heat of the Sun can cause cement to melt and bubble.) Anyone with basic skills can make a solar projection screen (see Chapter 3). If you do this, do remember to cap the finderscope securely. *Never* try to do solar projection with a catadioptric telescope, such as a Schmidt–Cassegrain or Maksutov, for the build-up of heat can damage it beyond repair.

## The Finder

The finder supplied with a 60mm is often a simple refractor, with a singlet objective lens and all-plastic optics, which is usually very easy to improve. Sometimes the simplest improvement is merely to remove the optics and use the remaining tube as a sighting device! These finders usually have a small diaphragm behind the objective whose purpose is to reduce the atrocious chromatic aberration of the singlet lens (see Fig. 1.2). They are quite effective at this, but they are also very effective at reducing the brightness if the image. If your finder has such a diaphragm, remove it – the resulting image will be much more colourful, but also much brighter.

If you want to replace the finder, either make one (junked binoculars again – see Fig. 1.3) or get one good

**Figure 1.2.** The 8mm diaphragm has only 1 mm more aperture than the eye of the likely user!

enough to use on the telescope you're saving up for – i.e. don't waste your money on a 30mm finder. If you do get, say, an 8 × 50 finder, you will notice that some deep-sky objects are brighter in the finder than in the telescope itself! Also, large objects of low surface-brightness, such as the Triangulum Galaxy (M33), are visible in the finder but invisible in the telescope itself, because the larger exit pupil will make images of extended objects brighter.

**Figure 1.3.** A 30mm finder made from a broken binocular.

Other options are to make simple "vee and blade" gunsight arrangements, a sighting tube, or a simple ×1 projected pinhole "red dot" finder. If you choose the latter option, you could use the existing finderscope tube to house it, and one of the lenses to project the pinhole image to infinity.

## Focuser

Some people suggest replacing the focuser with a 1.25-inch one. This might sound a good idea, but it is one of those areas where I think the money is better saved. You may find that a 1.25-inch option is available for your telescope, but otherwise a decent focuser, customised to your telescope, is going to be quite expensive; and the telescope will never give you those wide-angle deep-sky scenes, pregnant with faint fuzzies, which you hope to see. A hybrid diagonal may offer a better option, but, as noted above, there are potential pitfalls.

A possible exception to this rule is if you are upgrading a small reflector (e.g. a 114mm), where it is usually quite easy to change the existing focuser for a standard 1.25-inch one .

## Objective Lens

It comes as a surprise to most people that most of these telescopes have quite a good objective lens. This is therefore one of the last things to give your attention to, and then only if, as is usually the case, the internal surfaces are uncoated. Coated optics have a slight colour cast, usually blue, although green and red are becoming more common. Oiling a lens gives the least improvement for the most work. If you feel confident enough to tackle it (and I do not recommend this procedure unless you have some experience of tinkering with optics!), you may wish to see if oiling the lens gives any improvement. Most of these lenses are "contact doublets" which are in fact separated by three thin foil shims and whose internal surfaces are uncoated. This means that some light is lost through internal reflections. This light loss can be eliminated by replacing the air with oil.

First, dismantle the lens cell, making sure that you mark the lens elements for position, direction and orientation so that they can be reassembled exactly as they were. I use an HB (#2) pencil to mark the edges of a lens with a number indicating its position counting from the front (open end) of the telescope, and an arrow pointing to the front. The arrows line up on the lens elements. Remove and keep the three foil spacers, as you might wish to undo what you have done.

Inspect the internal surfaces of the lenses and, if they are coated, reassemble the lens in its cell *exactly* as it was previously, ensuring that you introduce no dirt, dust or fingerprints into the space between the lens elements. If you tilt the lenses as you put them back into the cell, they will probably jam, and may chip or break as you try to extricate them. The method I use is to support the lens from underneath, with a piece of clean lens tissue on my fingers, and let gravity do all the work as I gently allow the lens to lower into its cell, lifting it as soon as I detect even the slightest hint of binding. A common error at this stage is to over-tighten the retaining ring – tighten it snugly, then undo it a quarter of a turn – you should be able to feel a slight amount of movement of the lens as it is gently pushed.

The rear element is concave on the side that faces the front element. Carefully (it is made of flint glass, and scratches easily) place this element on a piece of lens tissue or acid-free paper, and place a drop of oil on it. I use thickened cedar oil, which is sold as microscope immersion oil, but others have used everything from cooking oil to engine oil. Place the front element on top of it (having checked that it is the right way round) and squeeze the two elements together, holding a tissue around their edges. The tissue will absorb most, but not all, of the excess oil that is squeezed out. Be careful – the resulting arrangement is *very* slippery! Line up the arrows and replace the lens in its cell. Tighten the retaining ring and carefully clean away the oil that oozes out. You will need to repeat this tighten-and-clean procedure daily for a few days. When no more oil is coming out, slacken the retaining ring off by a quarter of a turn and give the lens and cell a final careful clean. The remaining oil is held in place by capillary action. I did this to a 60mm lens over 5 years ago and it is still OK.

# Mechanical Upgrades

## Mount

The mount supplied with a 60mm is usually an insubstantial altazimuth mount which is not sturdy enough to stop the telescope shaking. You can increase its stability by suspending a weight from it. A plastic container, or even a bucket, suspended by a cord which runs through a hole in the tripod accessory tray, filled with a gallon or two (5–10 litres) of water will add a great deal of stability (see Fig. 1.4). There's not a lot else you can do about such mounts, except for regularly checking that all screws, bolts, etc. are snug and tight (see Fig. 1.5).

Equatorial mounts that come with these telescopes are of variable quality. Some, particularly those of 1960s vintage or older, are remarkably good for their

**Figure 1.4.** A 5-litre (1-gallon) container filled with water adds a great deal of stability.

**Figure 1.5.** All these nuts and bolts must be kept tight.

size (see Fig. 1.6); others are so flimsy that they have no business being anywhere near a scientific instrument. If yours is of the latter type, the best you can do is to use the water-container trick (above) and reduce telescope oscillation by other means, either by unbalanc-

**Figure 1.6.** This type of equatorial from the 1970s was of good, sturdy quality.

ing the axes of the mount (slip the telescope longitudinally to unbalance the declination axis; move the counterweight to unbalance the polar axis), or with a chain (see below), or both.

## Tripod

A good mount should be as steady as the Rock of Gibraltar. The chances are that yours isn't. If you have the necessary workshop skills, make yourself another tripod (and another altazimuth mount while you're at it). Alternatively, cut some trapeziums of thin plywood to fit between the upper parts of the tripod legs, and pin or tack them to the legs, as shown in Fig. 1.7. This has the disadvantage that the tripod doesn't collapse any more, and the accessory tray is nearly inaccessible. But a tripod is awkward to collapse or erect, with wing nuts getting lost in the grass; and filters don't fall out of a container in a coat pocket as easily as they do from an accessory tray, and neither do eyepieces dew up in your pocket. The modified tripod is much more stable, especially when the legs are not extended.

**Figure 1.7.** One of the trapeziums which add stability to the mount.

# Add a Chain

An insubstantial little tripod head doesn't do much to damp vibrations. One trick you might like to try is to hang a length of chain from the "big end" of the telescope (see Fig. 1.8). A 60cm (2-ft) length of chain gaffer-taped to the telescope is pretty good at damping vibrations, unless its frequency of oscillation resonates with that of the telescope, in which case shorten it by a link or two. Obviously it will oscillate in a breeze (but then, so does the telescope anyway), so you might like either to devise a way of making it easily detachable (e.g. add an eye to which it clips) or to fix it to the telescope (e.g. with another length of gaffer tape) .

Finally, when you come to replace your 60mm with something bigger and better, don't chuck it out or sell it – it still has potential. Its small size means that it will fit easily into a small car, so you can always have a telescope to hand. With a little ingenuity you can make a small mount that fits onto the car. There are also mounts available for fitting such a telescope to the car

**Figure 1.8.** A length of chain damps vibrations.

**Figure 1.9.** The 60mm refractor used as a guidescope.

window, so the observer can observe from the comfort of the car seat. Alternatively, the 60mm can become your first guidescope once you have started doing guided astrophotography (see Fig. 1.9).

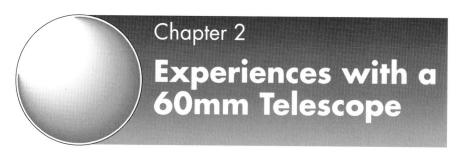

# Chapter 2

# Experiences with a 60mm Telescope

Robert Hatch

"I've found Neptune!"

"How do you know it's Neptune?"

"Well, it moves from night to night against the background sky."

"Can I have a look? – Oh!"

That, I hear, was the gist of a conversation between an astronomical friend of mine (the one who found Neptune) and another, more prominent astronomer (who questioned his finding) at a star party some years ago. I understand that the experience changed the attitude of the latter to small telescopes – in this case a 100mm (4-inch) refractor, I recall – for ever. He was "converted". This anecdote impressed me also, for it shows what can be done with a small telescope.

I have been passionate about astronomy from my early teens, and my first telescope, a 60mm (2.4-inch) refractor with a focal length of 910 mm was bought from a college friend in the early 1970s. It is mounted on a (fairly) sturdy tripod, whose main brace doubles as a handy eyepiece holder, and has a selection of eyepieces: 20 mm, 9 mm and 4 mm and a ×2 Barlow lens. I soon found that one needs to be aware of the shortcomings of these lenses. In particular, the 4mm with the Barlow gives great magnification (×455!) but absolutely useless resolution – in fact, to all intents and purposes it's unusable. This is a classic caveat for the smaller (cheaper?) telescope.

The 20mm has good eye relief but a small image scale; even so, it is ideal for observing mutual phenomena of Jupiter's Galilean satellites, for example, yielding a magnification of ×45 or so, and lunar occultations (see Chapter 3). So, despite the telescope's small size and relatively large focal ratio (*f*/15), it is useful for certain types of observation. The 9mm eyepiece gives a better image size and is more suited to studying planetary detail. But again, the quality can be questionable, especially when using the ×2 Barlow; Barlow lenses need to be of good quality to get the best from an eyepiece. A good clear night with good seeing is certainly called for.

The large focal ratio gives the scope a small field of view, depending on the eyepiece used – typically just over half a degree with the 20mm, ideal for a full view of the Moon. This is where the small finderscope comes into its own, having as it does a magnification of ×6, but more importantly a much larger field of view, at least 5° or so.

In later years I acquired a 300mm (12-inch) *f*/6 reflector on an altazimuth mounting, but this is hardly portable, at least at the time of writing (I have plans to replace the tube with a lighter structure), so I have kept

**Figure 2.1.** Telescope on tripod with the mounted camera and 200mm lens.

the 60mm. It is handy for field trips, as it conveniently detaches from its collapsible tripod, and can be used both as a telescope and as a platform for guided photography. This second function is the main theme of this chapter. Using a suitable attachment, a camera can be piggybacked to the telescope as shown in Fig. 2.1; Fig. 2.2 shows a close-up of this arrangement. The mount is more versatile than the usual single ball-joint photographic mount, allowing much greater freedom of movement for the camera.

When I first bought the 60mm telescope, it had fitted to it a small mains-synchronous motor (powered from the domestic supply) adapted by the original owner from an electric meter. It worked well, but, over time, the wiring became suspect and I discontinued its use for safety reasons, preferring to use

**Figure 2.2.** Close-up of the camera mounting bracket.

the slow-motion hand controls in the absence of any electrical drive.

Whether or not a mains- or battery-powered system is employed is up to the user. Both are available. For safety, a mains-powered system needs to be protected by an isolating transformer, to avoid having a direct connection to the mains supply, to reduce the risk of shock in the event of a fault developing in the wiring. If the telescope is to be used in the field, away from a mains source, there are two options: either use an "inverter", which enables, say, a car battery to generate the required voltage, or fit the telescope's RA drive with a low-voltage d.c. system, which will run from a small 9V pack, for example.

Computerised accessories are also available, which control both the RA and declination axes, but these are specialist devices at specialist prices!

The apparent daily revolution of the night sky around the Earth (caused, of course, by the Earth rotating on its axis) takes 23h 56m 4.099s, to be precise. This period is called the sidereal day, meaning that it is measured "fixed star to fixed star". It differs from the (mean) solar day (of 24 hours) by just under 4 minutes, as the solar day is measured "mean sun to mean sun". The Earth moves nearly 1° in its orbit around the Sun during the course of a day, so this 4 minutes is the time taken for the Earth to "catch up" with the Sun again. This difference need not concern us for our purposes, but is mentioned here as a reminder to the beginner. For really accurate long exposures, on faint, telescopic objects, it would become important. Figure 2.3 shows how a telescope, set up at a given latitude, is arranged to counteract the rotation of the Earth. Objects can then be kept in the field of view of the telescope as the Earth rotates, and star trails in photographs of more than a few minutes' exposure are eliminated.

The other essential is to align the polar axis. The best way to do this is to sight along this axis with it pointing at the pole star, with the elevation axis loosened (be careful to support the telescope, as it will now droop on this axis under its own weight) with the azimuth axis free to rotate on the mounting. The pole star will then be hidden by the azimuth axis when the eye is "lined up" to view along this axis, but a slight movement of the head from side to side to check the position of the pole star should enable you to align the axes sufficiently well to take photographs by the piggyback method. The process is completed by clamping the

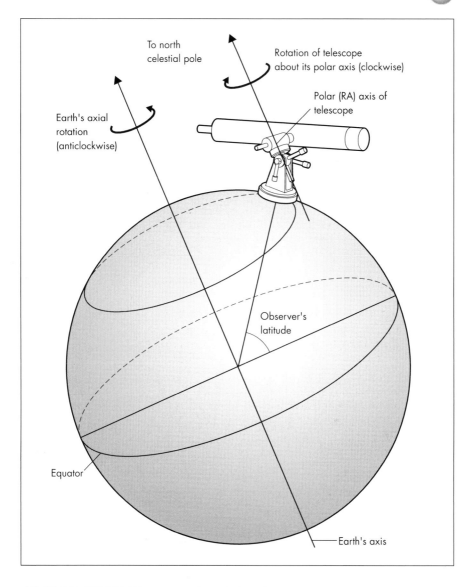

To north
celestial pole

Rotation of telescope
about its polar axis (clockwise)

Polar (RA) axis of
telescope

Earth's axial
rotation
(anticlockwise)

Observer's
latitude

Equator

Earth's axis

**Figure 2.3.** The Earth's axial is countered by the rotation of the right ascension axis of an equatorially mounted telescope.

azimuth and altitude axes. With some instruments it is possible to physically "sight through" the mounting by removing some fittings; the pole star (Polaris) would then be visible through the axis mount, which may make things easier and improve the accuracy of alignment. Alternatively, one could use a protractor on which one's latitude is marked, accurately aligned and firmly fixed to the telescope mounting.

Reference is made in the above paragraph to the azimuth and elevation axes. The equatorial head is in

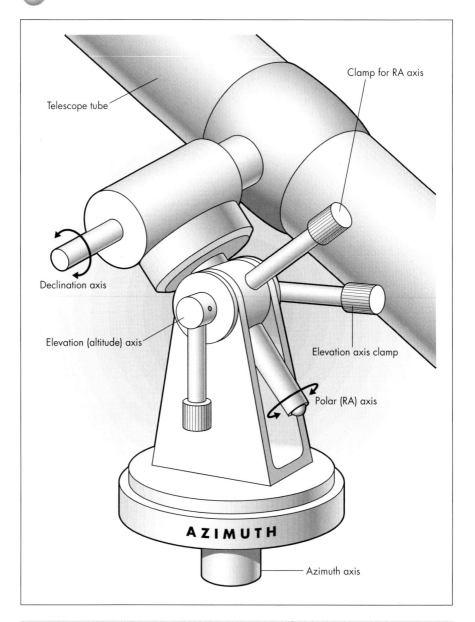

**Figure 2.4.** A typical equatorial head. The elevation axis is fixed at an angle equal to the observer's latitude (which is also the altitude of the north celestial pole above the observer's horizon). The azimuth clamps are used to fix the RA axis in the north–south direction.

fact a special case of the altazimuth mounting. The head is mounted on an altazimuth, but the RA and declination axes are functions of the Equatorial head itself and enable the telescope to track parallel to the Earth's motion. The elevation (altitude) axis is fixed at an inclination equal to the latitude of the observer, and is pointed at the pole star by fixing the azimuth axis in the north–south direction. Figure 2.4 shows the relationship between the two sets of axes.

After carefully aligning the equatorial axis as described, a camera can be mounted piggyback by a suitable mounting bracket (Fig. 2.2 shows a close-up of the author's fitting), with the appropriate lens for the required shot (Fig. 2.1 shows a 200mm f/3.5 lens). Then carefully align the camera on the required starfield. If the telescope is fitted with a motorised drive, switch the motor on. If not, the telescope itself can be used to follow the diurnal motion of the sky by focusing it on a suitable star and using hand guiding. You may need to realign the camera onto the selected starfield by adjusting its orientation on the mounting bracket. Note also that it is not necessary for the telescope to be pointing in the same direction as the camera. It may sometimes be desirable to guide "off-axis", especially where the telescope tube may obstruct the camera's field of view.

The camera can be fitted with a Bowden cable shutter release to make operation as jog-free as possible. This cable has a small locking screw to lock the shutter open in the B (for "bulb"). An idea popular among amateur astrophotographers is what has been called the top hat trick. One places a hat or, more usually, a piece of black cloth, over the objective lens before opening the shutter. If hand guiding is to be used, then proceed as follows:

- Select a guide star, and centre it in the eyepiece.
- Open the shutter.
- Remove the cloth carefully, then commence hand guiding, keeping the guide star in the eyepiece and timing the exposure (see below for timing).

It is important to keep the guide near the centre the field of view for the duration of the exposure, using the slow-motion hand drives, otherwise blurring will become evident on the final image. Compare the two photos of Orion: Fig. 2.5a shows the stars trailed (no guiding), while Fig. 2.5b was obtained with hand guiding. Both exposures were for exactly 5 minutes, on ISO 100 Ilford Delta Professional black and white film.

Finishing the exposure is the reverse of the initial procedure: cover the objective lens with the cloth, then close the shutter.

a

b

**Figure 2.5.** Five-minute exposures of Orion with **a** no guiding (star trails) and **b** hand guiding (sharp star points).

Timing can be achieved by whatever means is most convenient. A friend of mine uses a large clock which "ticks" away the seconds, but a clock with a clear red digital display could also be used, although you have to keep one eye on it. With a little practice it is possible to keep one eye on the clock while keeping the guide star in view. In fact, a movement of the hand guider every 10 seconds or so is usually sufficient. It is very much up to you how you manage the timing. Perhaps get someone else to do the timing for you, engaging others in the activity. If other people likely to be moving about in the vicinity of your camera, as at a star party for example, then make sure that no one walks into the tripod mount, and dissuade them from flashing torches around, even red ones!

Choose the camera lens carefully. The support shown in Fig. 2.2 is not the most stable type. With heavy or long lenses the camera assembly could shake, but in practice lenses of up to 200 mm are usable with no problems. I have tried a 500mm, but with difficulty; there is always the increased risk of shaking – and thus of blurred images – with heavier, more unwieldy lenses. Clearly, the sturdiness of the tripod is also a factor, so check beforehand that the legs are clamped as tightly as possible to the upper bearing plate assembly, and that the set-up is as stable as possible.

The 60mm refractor is very well suited to projecting an image of the Sun, and mine was originally supplied with a small solar projection screen which could be attached to the tube. (Catadioptric telescopes should *never* be used for solar projection work, as they can be seriously damaged by the build-up of heat.) The different elements of the lenses are air-spaced, are so not subject to damage by the concentrated heat of the Sun's rays as are the multi-element, cemented structures. One thing to note here is that, if you use a right-angle prism (star diagonal), the image of the Sun will be laterally inverted; *in addition* to the inversion of the image as given by any astronomical telescope (unless an erecting lens or prism, is provided for terrestrial viewing). Also, the star diagonal will add approximately 60 mm to the focal length, so the focuser has to be racked in by this much to achieve correct focusing. Great care should be taken when using this prism for solar work, especially where there is a gathering of people, that it is not rotated to enable folk to view directly! The idea of it, after all, is to add convenience to the telescope's (night) use, especially when observing objects towards the zenith.

I have discussed the use of this prism with the solar projection screen. One important caveat cannot be overemphasised at this point: *never* be tempted to use the eyepiece solar "filter" which comes with your telescope – it is likely to shatter in the eyepiece, where it is intended to be fitted, with obvious dire consequences (see Chapter 1). The only filter that should be used for viewing the Sun is a properly fitting, full-aperture solar filter. The power of the Sun's rays can easily be demonstrated. One friend puts it this way, "It'll bore a hole straight through ya head!" – but then he does have a way with words. He's got a point though: just *don't* let it be a point of dead, heat-scarred retina that will *never* see again.

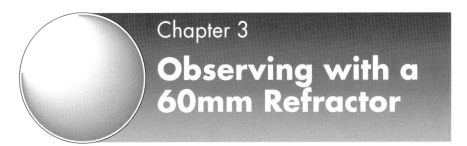

# Chapter 3
# Observing with a 60mm Refractor

Stephen Tonkin

The 60mm (2.4-inch) refractor is a much-maligned telescope. The major problem with this class of telescope is that purchasers have expectations of it which vastly exceed its capabilities. This is usually a consequence of the advertising, some of which can be extremely misleading. The telescope usually comes as either a *f*/15 or an *f*/11.2 model. The *f*/15 models generally have slightly better objective lenses, giving far less false colour. Mine is mounted on a sturdy equatorial, giving it the advantage that movement in one axis only is necessary – a distinct advantage at high magnification. Used within its capabilities, the telescope has potential as an amateur instrument, particularly in two areas: solar observation and lunar occultations.

## Solar Observation

The safest way to observe the Sun is to project its image onto a screen. It is a simple matter to concoct a solar projection fitting for a 60mm telescope using simple tools and readily available materials. The projection attachment shown in Fig. 3.1 comprises a plywood board attached to a length of aluminium angle, which is itself bent so that the centre of the board is coincident with and perpendicular to the optical axis of the telescope. The angle is attached to the telescope with a Jubilee clip (hose-clip). A sheet of paper (or a Stoneyhurst disc) is held onto the board by a pair of

**Figure 3.1.** The arrangement for solar projection.

plastic slide binders. A collar on the telescope shades the screen (and the finder) from direct sunlight. This does not give as much contrast as does a fully enclosed projection system, but it has three distinct advantages:

1. It is simple and quick to make, requiring only rudimentary workshop skills.
2. It is extremely light, which makes it easily portable and places less demand on the telescope mount.
3. The solar image can be viewed by several people at once, making it suitable for public "Sun parties" and for school demonstrations (see Fig. 3.2).

Some thought needs to be given to the eyepiece used for projection. Some observers have noted that solar projection can heat cemented lenses in the eyepiece to the extent that the cement crazes. The obvious solution is to use an uncemented eyepiece such as a Huygenian (with which these telescopes are often supplied) or a Ramsden. I use a 20mm Huygenian that I keep purely for this purpose.

**Figure 3.2.** The open screen allows easy viewing of solar images, as with this partial eclipse.

There is often insufficient focal range with these telescopes to permit the eyepiece to focus an image of the Sun on the screen. Probably the simplest solution is to make an extension tube by adapting the Barlow lens or the image erector with which these telescopes are usually supplied, and which are usually of such poor quality that they are all but useless for their intended purpose.

## Lunar Occultations

From time to time, solar system bodies occult one another or a star. There are essentially six categories of occultation:

1. *Occultation of stars by asteroids.* Owing to the rarity of occultations of bright stars by asteroids (normally observed as a dimming or extinction of the star as the asteroid passes in front of it) from any given location, it is probably not worth keeping a 60mm just for this purpose. Occultations of stars can be observed – provided, of course, that the occulted star is sufficiently bright to be seen through the telescope in the first place. However, if the data obtained are to

be of use in determining the shapes of asteroids (the primary purpose of such observations), photoelectric recording of the observation is necessary, and this is something to which the common 60mm refractor is singularly unsuited!

2. *Occultation of stars by planets.* These can provide information about a planet's rings or atmosphere, but again, photoelectric recording is necessary if the data are to be useful.

3. *Occultation of moons by the parent planet.* Observations of these are fraught with difficulties because of the small size of the moons' discs and the effect of the planet's atmosphere, which combine to produce a slow fading which usually cannot be timed properly.

4. *Mutual occultation of a planet's (usually Jupiter's) moons.* These are certainly observable with a 60mm refractor, but the data are unlikely to be useful other than for personal interest.

5. *Occultation of a planet by the Moon.* These rare but intriguing events can be observed for interest.

6. *Occultation of stars by the Moon.* These are fairly common events whose timings are of scientific value, and are something to which the 60mm refractor is well suited, and it is on these that we shall concentrate.

There are two further subdivisions for lunar occultations:

1. *Graze occultations.* These are multiple disappearances and reappearances (immersions and emersions) of a star behind the limb of the Moon near one of its poles. They are visible only over a narrow track of land (the "graze track"), which is typically about a kilometre (about half a mile) wide. Owing to the intricacies of recording these events, they are ideally undertaken as group activities co-ordinated by a local astronomical society or club, with an experienced member in each observing team.

2. *Total occultations.* These are an immersion (disappearance) or an emersion (reappearance) of a star at the lunar limb. Immersions, particularly if the occultation is by the dark limb of the Moon, as opposed to the sunlit limb, are the easiest to see. The narrow field of view of the 60mm refractor makes it easier to concentrate on the star without the distraction of the illuminated part of the Moon. They are relatively easy for a lone observer to observe and record.

# Preparation

The main source of predictions is the International Occultation Timing Association (IOTA), which is also the recognised co-ordinating body for lunar occultation observations. Its predictions are usually disseminated by national astronomical organisations, and are also available on the internet (http://www.lunar-occultations.com/iota/iotandx.htm) . Also available from IOTA is Dave Herald's lunar occultation software, OCCULT, which generates predictions and report forms (see Fig. 3.3). Using good-quality planetarium software, it is also possible to generate one's own predictions merely by observing which stars the Moon occults during a fast-motion simulation. For a beginner this also has the advantage that it generates a view of what you may expect to see, especially if you set the orientation and field of the screen image so that it corresponds to what you see in the eyepiece (see Fig. 3.4).

Try to choose an observing site that is protected from the wind, and away from any objects such as trees or telephone poles that could interfere with the occultation. In order for the timing to have scientific value, you need to know the location of your observing site to a precision of 25 metres (80 ft), and preferably 10 metres (30 ft). This can be obtained from large-scale maps or from Websites such as http://www.mapblast.com. Alternatively, you may use a GPS (Global Positioning System) receiver if you have one whose precision is sufficiently reliable.

**Figure 3.3.** The OCCULT software generates prediction details.

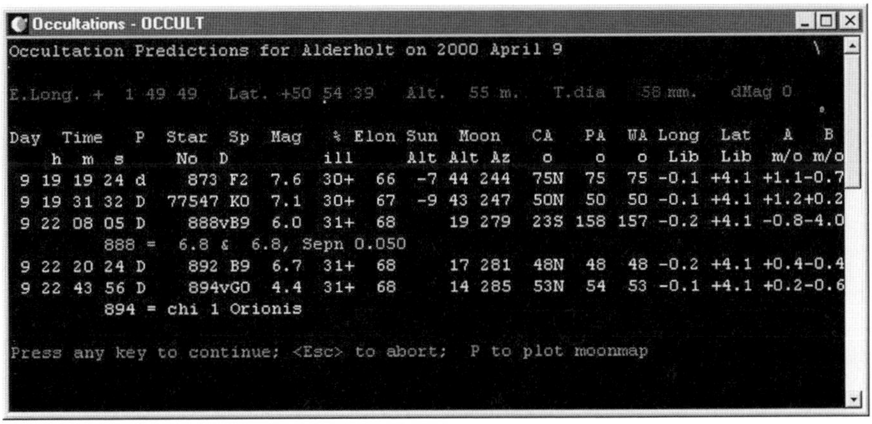

```
Occultations - OCCULT                                              _ □ X
Occultation Predictions for Alderholt on 2000 April 9                  \

E.Long. +  1 49 49  Lat. +50 54 39  Alt.  55 m.  T.dia  58 mm.  dMag 0
                                                                        o
Day  Time  P  Star  Sp  Mag  % Elon Sun  Moon   CA   PA  WA Long  Lat   A    B
     h  m  s      No  D      ill     Alt Alt Az  o    o   o  Lib  Lib  m/o m/o
9 19 19 24 d    873 F2  7.6  30+ 66  -7 44 244  75N  75  75 -0.1 +4.1 +1.1-0.7
9 19 31 32 D  77547 KO  7.1  30+ 67  -9 43 247  50N  50  50 -0.1 +4.1 +1.2+0.2
9 22 08 05 D    888vB9  6.0  31+ 68     19 279  23S 158 157 -0.2 +4.1 -0.8-4.0
             888 =  6.8 & 6.8, Sepn 0.050
9 22 20 24 D    892 B9  6.7  31+ 68     17 281  48N  48  48 -0.2 +4.1 +0.4-0.4
9 22 43 56 D    894vGO  4.4  31+ 68     14 285  53N  54  53 -0.1 +4.1 +0.2-0.6
             894 = chi 1 Orionis

Press any key to continue;  <Esc> to abort;  P to plot moonmap
```

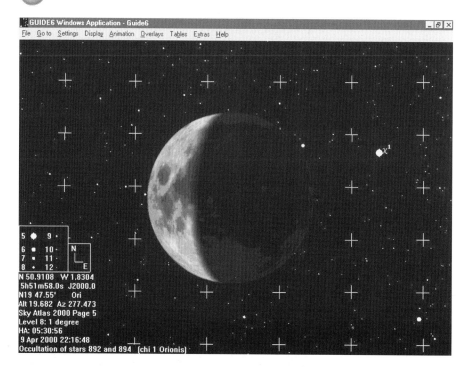

Within the image:

GUIDE6 Windows Application - Guide6

File  Go to  Settings  Display  Animation  Overlays  Tables  Extras  Help

5  ◆  9 ·
6  ●  10 ·
7  ■  11 ·
8  ·  12
N 50.9108  W 1.8304
5h51m58.0s  J2000.0
N19 47.55'  Ori
Alt 19.682  Az 277.473
Sky Atlas 2000 Page 5
Level 8: 1 degree
HA: 05:30:56
9 Apr 2000 22:16:48
Occultation of stars 892 and 894  (chi 1 Orionis)

χ¹

# Timing

There are various methods of timing occultations, for all of which you need to calibrate your timing against a standard timebase, such as radio time signals (*not* time announcements, which are notoriously inaccurate!) or the telephone time service. If you are using the latter, it is preferable to do it via landline because there is usually a small delay when this is transmitted over mobile phones. It is also possible to obtain clocks that are synchronised by radio time signals. If you use one of these, do check that it is accurate between the radio corrections – some of these clocks rely on the radio correction to compensate for poor timekeeping between corrections.

You also need to ascertain your personal equation – the delay between the occurrence of an event and your operating the timing mechanism (usually a stopwatch). Perhaps the simplest way to do this is to use software designed for the purpose, in which a star's image on a screen disappears and reappears at random intervals, and your response time is then evaluated. It is important to try to emulate the conditions that you will encounter at the eyepiece when you use this software –

**Figure 3.4.**
Planetarium software can give a preview of what will be seen. Generated by Project Pluto's GUIDE v6.0

**Table 3.1.** Reduction of timings

| Event | Stopwatch | Clock time |
|---|---|---|
| Start timing | 0h 0m 00.0s | 22h 35m 00.0s |
| Immersion of star (Lap #0) $(y)$ | 0h 8m 46.2s | |
| | | |
| Lap #1 $(z_1)$ | 0h 13m 0.1s | 22h 48m 00.0s |
| Lap #2 $(z_2)$ | 0h 14m 0.0s | 22h 49m 00.0s $(t)$ |
| Lap #3 $(z_3)$ | 0h 15m 0.0s | 22h 50m 00.0s |
| Personal equation $e = 0.3$s | | |
| | | |
| *Calculation* | | |
| Mean difference | $a = z - y = 14$m 0.0s | |
| Difference | $b = a - y = 5$m 13.8s | |
| Subtract personal equation | $x = b - e = 5$m 13.5s | |
| Time of event | $t - x = 22$h 43m 46.5s | |

there is a world of difference between a comfortable seat in a warm room and an awkward stance at a telescope eyepiece on a night so cold that your observing eye is streaming tears!

When it comes to the process of timing, I favour simplicity. I use a digital stopwatch with a ten-lap facility. I start the watch, then take the first lap time from the telephone time service, and note the time. I use the lap facility to record the occultation or occultations, then return to the telephone time service and take three more times from it. The last three times serve as a standard time reference, and can be averaged. This obviates the need to apply a personal equation: synchronising the button-pressing with the regular "pips", which are a second or so apart, reduces the personal equation to approximately zero.. By taking times before and after the occultation, any inaccuracy in the stopwatch, such as can result from a weak battery or extreme cold, will be revealed. There is less opportunity for miscalculation, and more chance of detecting a faulty stopwatch, if all the telephone time service times are taken at the start of a whole minute , and are reduced (analysed) as shown in Table 3.1.

# Reporting Observations

Accurate occultation timings are of scientific value: the occultation of stars by the Moon allows us to improve of our knowledge of its orbit, while the occultation of stars by asteroids permits the size and profile of the occulting body to be better determined. The International Lunar Occultation Centre (ILOC) in Tokyo, Japan is the recognised centre for processing occultation data. Report forms can be obtained from ILOC or, if you prefer, can be generated by the OCCULT program. When you report your first occultation you will be allocated an occultation station designation.

## Conclusion

Lunar occultations are fascinating events in their own right, but the sense that one is contributing to scientific endeavour makes occultation timing an especially fulfilling activity. It is even more satisfying to know that one has made this contribution with an telescope that is often denigrated as being unworthy of being called a scientific instrument!

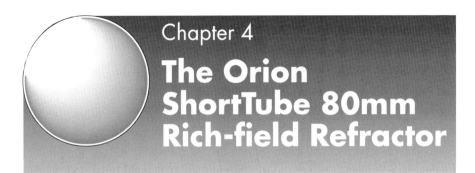

Chapter 4

# The Orion ShortTube 80mm Rich-field Refractor

Dave Mitsky

## Introduction

Novice shoppers face a bewildering variety of choices in today's telescope market. Some prospective buyers may opt for a fairly substantial Newtonian reflector on an altazimuth Dobsonian or equatorial mount, or perhaps a fork-mounted 125mm (5-inch) or 200mm (8-inch) Schmidt–Cassegrain catadioptric. However, the sheer bulk of such telescopes and time required to set them up and to allow for adequate athermalization (i.e., reaching thermal equilibrium with the ambient temperature) means that there will be many occasions when they are not used simply because of a lack of time or energy. The sexy 90mm (3.5-inch) and 125mm Maksutov–Cassegrains will catch a few eyes, but the limited fields of view produced by their high focal ratios and the less than ideal finderscopes sometimes supplied with them make it unlikely that newcomers to astronomy will have much success in using them to locate any but the easiest of celestial targets. (Those models equipped with robotic GOTO capability obviously do not suffer from this limitation.) Others may want to test the waters of amateur astronomy in a gentler fashion by buying a small first telescope, say a 60mm (2.4-inch) refractor or a 114mm (4.5-inch) Newtonian. These newcomers are primarily interested in an instrument that is inexpensive and simple to use. Unfortunately the quality of many such telescopes is

questionable, particularly in the stability of the
mounts. Nature-lovers are looking for a telescope that
can be used for bird-watching and daytime terrestrial
viewing. Spotting scopes perform this function well
enough, but leave much to be desired as astronomical
telescopes. And then there are those who wish to purchase a telescope that is extremely portable, one that
can be carried in a backpack while camping or taken
on board a commercial airliner. Here the requirements
are a diminutive size and a manageable weight.

Is there a telescope that can serve the interests of
these disparate users without causing their bank balances to plummet or their credit cards to melt down?
The answer is a resounding "yes." (Perhaps it would be
more accurate to say, at the very least, a qualified "yes,"
as I shall explain.) In recent years a flood of inexpensive Chinese 80mm $f/5$ achromatic refractors[1] has inun-
dated the astronomical landscape. Sold under various
names and with differing standard equipment, these
telescopes provide very satisfying low-power views
and, when "tweaked" (i.e., modified) and used with
high-quality accessories, they can be surprisingly good
performers when it comes to lunar and planetary
observing. However, one can hardly expect sterling
planetary performance from such short focal length
achromats, which excel primarily as rich-field instruments.[2] Nevertheless, in my opinion, these small
refracting telescopes make far better first telescopes
than the typical department store offerings.

Seemingly clones of the Japanese Vixen New Planet
80S, these refractors are available from a number of
vendors under such names as the Orion ShortTube 80,
the Astrovisioneer, the Bauer Optik ST80, the Celestron
Firstscope 80 EQ Wide Angle, the Helios Startravel-
80T, the Konus Vista-80, the Bresser Champ, the Sky
Instruments Sky Watcher SW 804 (formerly the Vista
508), and the teleHOON RF80F5A1. To the best of my
knowledge a Chinese company by the name of Synta is
responsible for manufacturing all these optical instru-

---

[1] An achromatic refractor is one in which the lens focuses
most of the spectrum, except for the extreme short and long
ends, to nearly the same focal point.
[2] A rich-field (or richest-field) telescope is a small-aperture
instrument of short focal length that produces a very wide
field of view. The term "rich field" refers to the fact that such
a telescope displays the greatest number of stars possible in a
single field of view.

ments. Although the ShortTube 80 and its relatives may differ in tube color (white, black, blue, and orange) and supplied accessories (eyepieces, diagonals, and mounts), they are all air-spaced doublet refractors with clear apertures of 80 mm (3.1 inches) and focal lengths of 400 mm (15.7 inches).

Since the ShortTube 80 and its siblings are two-lens element achromats of short focal length, the bane of refracting telescopes – chromatic aberration, or false color – rears its ugly head when moderately high magnifications are attempted on objects such as bright stars, Jupiter, Venus, or the Moon.

## The Orion ShortTube 80

The version that Orion Telescope & Binoculars[3] currently sells for US$199.00 (it was originally offered for $249.00, and was later reduced to $229.00) has an optical tube that is made of aluminum, and is painted white.[4] Included are a 90mm (3.5-inch) aluminum dew/glare shield, an end cap that can double as an aperture mask, a functional $6 \times 30$ achromatic finderscope that is mounted on a dovetail bracket, a 45° erect-image (Orion uses the term "correct-image") prism diagonal, and a standard $\frac{1}{4}$-inch – 20tpi mounting plate (see Fig. 4.1). Unfortunately the mounting plate is made of plastic, which raises concerns about its durability. Two Kellner oculars are supplied: A 25mm that yields a magnification of ×16 and a 10mm that produces ×40. The objective lens of my ShortTube 80 has an anti-reflection multi-coating that is green in color. The optical tube is baffled to reduce internal reflections. The ShortTube weighs in at 1.7 kg (3 lb 13 oz) and is 39.4 cm (15.5 inches) in length with the dew shield removed. The one-page instruction sheet that Orion supplies with the ShortTube 80 leaves something to be desired, particularly for first time telescope users.

The finderscope, which is stopped down to an equivalent of 24 mm in order to reduce chromatic aberration, is held in place by a rubber O-ring and three

---

[3] http://www.telescope.com/interact/item.asp?itemno=A021
[4] A number of reports that have appeared on the Internet indicate that there have been minor variations in the construction of the ShortTube 80 since its introduction.

**Figure 4.1.** The Orion ShortTube 80mm rich-field refractor, 90° mirror diagonal, and 8mm Tele Vue Radian eyepiece mounted on a Bogen 3011 tripod and 3030 head.

locking screws. (I prefer a six-lockscrew finderscope bracket.) Loosening the objective end cap, setting a locking ring to an appropriate position, and then tightening the end cap will bring the finderscope to focus. The ShortTube 80's focuser is of the rack-and-pinion variety and works quite smoothly, having approximately 8.9 cm (3.5 inches) of back-travel. The knobs of the focuser are large enough to be turned easily when wearing gloves. I sometimes wish that the focuser set screw would grip the diagonal a little more firmly.

The amateur astronomical community has received the ShortTube 80 most favorably. *Sky & Telescope* magazine published a positive review in the March 1997 issue. A version of this article is available on the Internet.[5] Internet telescope reviewers Todd Gross[6] and Ed Ting[7] both praise the instrument. In fact, Mr. Gross considers the ShortTube 80 to be a best buy. Rod Mollise, the founder of email list 80f5, has posted his experiences with his Celestron 80 EQ WA at the Cloudy Nights Telescope Reviews home page.[8]

---

[5] http://www.skypub.com/resources/testreports/telescopes/richfield.html
[6] http://www.weatherman.com
[7] http://www.scopereviews.com
[8] http://www.cloudynights.com/reviews/st80.htm

# Capabilities

A telescope with an aperture of 80 mm has a resolving power of about 1.4 arc seconds and should be capable of revealing stars of approximately 12th magnitude. The ShortTube 80 is far from the ideal planetary telescope, but it nevertheless performs much better than one might expect when it comes to the planets. Saturn's rings and its brightest moon, Titan, can be seen quite easily. Three other moons – Dione, Rhea, and Tethys – are also within the capabilities of this instrument. When the conditions for observing are good it is possible to detect the Cassini Division and banding on the planet's disk as well. Jupiter's major cloud belts and several lesser ones are discernible. Reports of seeing the Great Red Spot, Galilean shadow transits, and even the satellite transits themselves have surfaced on a new email list called 80f5,[9] which is dedicated to discussing 80mm $f$/5 rich-field refractors.

Some of the best double stars in the heavens can be split (i.e., resolved) with the ShortTube 80. Among them are Castor ($\alpha$ Geminorum), Albireo ($\beta$ Cygni), Rigel ($\beta$ Orionis), the famous Double Double ($\varepsilon$ Lyrae), Almach ($\gamma$ Andromedae), Algieba ($\gamma$ Leonis), and Mizar ($\zeta$ Ursae Majoris). Englishman Dave Randell has reported resolving two close binaries with his modified Konus Vista-80: Izar ($\varepsilon$ Boötis) and Alula Australis ($\xi$ Ursae Majoris), the latter being quite an accomplishment for such an instrument, and others have split $\gamma$ Virginis using ShortTubes.

In the realm of the deep sky, all of the Messier objects and numerous objects from the *New General Catalogue* (NGC), the *Index Catalogue* (IC), and other astronomical catalogs lie within the capabilities of the ShortTube 80, although many will be far from impressive in appearance through an 80mm telescope. The strong suit of the ShortTube 80 is the wide field of view made possible by its fast focal ratio and small aperture. A 32mm Plössl with a field stop of 27 mm will yield a true field of view of almost 4.0°, the maximum possible in this telescope using a 1.25-inch eyepiece. This is more than enough to frame almost any large deep-sky object. It is quite possible to see the Andromeda Galaxy (M31), the Pleiades (M45, see Fig. 4.2), the Coathanger

[9] http://www.egroups.com/community/80f5

(Collinder 399, see Fig. 4.3), the Cygnus Loop or Veil Nebula (NGC 6960, NGC 6979, NGC 6692, NGC 6995), and the North America Nebula (NGC 7000) in their entirety through a ShortTube 80.

The ShortTube 80 has a number of other uses:

- It can be used quite effectively as a spotting telescope for bird watching and other activities. The supplied 45° prism diagonal will provide the proper perspective.
- This telescope also serves quite well as a 400mm $f/5$ telephoto lens. I have obtained rather good short-exposure astrophotographs from a stationary tripod, and longer exposures with the ShortTube 80 piggy-backed on telescopes equipped with clock drives. Tripod-mounted exposures must be kept quite short, no more than a total of 10 seconds when photographing near the celestial poles and only 2.5 seconds at the celestial equator, to avoid star trailing. Employing the ShortTube 80 as a telephoto lens requires a threaded extension tube to move the focus to about 4.6 m (15 ft), otherwise it will focus at approximately 30.5 m (100 ft). When used for prime focus astrophotography only a T-ring, which mates the camera to the optical tube, is needed. The T-ring is screwed directly onto the end of the threaded focusing tube.

**Figure 4.2.** Venus and the Pleiades (M45) photographed on April 4, 1999, 5-second exposure, Pentax K1000 on a stationary tripod, Fuji Super G Plus 400 film.

**Figure 4.3.** The Coathanger (Collinder 399): a 2-minute piggybacked exposure on Kodak Gold Max 400 film, Pentax K1000, September 11, 1999.

- For eyepiece projection photography the extension tube, a universal camera adapter, and an eyepiece will be needed.

- Large telescope owners who are searching for a substantial finderscope will do well to consider the ShortTube. The capability to change focus easily, to interchange eyepieces and filters, and to use an Amici prism to produce erect-image views is advantageous in a finderscope.

- Although I have seen the ShortTube 80 function as a CCD autoguider guidescope, it is likely that it will be inadequate for general use as an astrophotography guidescope.

## Tripods and Mounts

At just under 1.8 kg (4 lb), the ShortTube 80 is light enough to be used in an altazimuth mode on a number of different photographic tripods. Orion recommends the Paragon HD-F2 for $109.00. During the summer of 2000 the Orion catalog added another option – the ShortTube 80 EQ refractor, complete with an Orion EQ-1 equatorial mount, for $289.00. Of course, one can also decide to mount the ShortTube 80 on a more elaborate mount such as the Orion AZ-3 altazimuth mount ($169.00), the EQ-2 equatorial mount ($149.00), or the SkyView Deluxe EQ mount ($239.00). $\frac{1}{4}$-inch -20 ($\frac{1}{4}$-inch UNC thread) adapters ($9.95) are required for the latter two mounts.

It has been suggested that the Celestron #91503 CG-3 equatorial mount[10] would also be a good match for the ShortTube. Canadian amateur astronomer Geoff Gaherty reported on the Talking Telescopes email list[11] that he has achieved very satisfactory results using a Tele Vue Up-Swing head[12] and a heavy-duty Manfrotto tripod. Matt Tarlach of California has said that the Tele Vue Tele-Pod head[13] mounted on a Bogen 3221 tripod is also an excellent combination.[14]

When I purchased my ShortTube 80 I already owned a Bogen (Manfrotto) 3011 video tripod. However, I replaced the original tripod head with a Bogen 3030 quick-release head for the sake of convenience and to save wear and tear on the mounting plate. (A fluid pan head such as the Bogen 3126 might be a better choice for easy tracking.) The Bogen 3011 handles the weight of the telescope with ease, as does the 3211, which is merely a black version of the chrome 3011. It has been reported that Bogen model 3001 (3205) and 3021 (3221) tripods are also compatible with the ShortTube 80. The Bogen 3270 (410) is a geared tripod head with slow motion controls that has been reported to work quite effectively with the ShortTube; the drawback is that it is quite expensive.[15]

# Chromatic Aberration and Maximum Magnification

Chromatic aberration, or false color, is the result of various wavelengths of light focusing at different distances from the objective lens of a refracting telescope. It is manifest primarily as a purple colored fringe or

---

[10] http://www.celestron.com/access/eqm.htm
[11] http://www.egroups.com/message/telescopes
[12] http://www.rahul.net/amall2/tvo/pg22.htm
[13] http://www.rahul.net/amall2/tvo/tele-pod.html
[14] http://www.egroups.com/message/80f5/1218?&start=1199
[15] Information on Bogen tripods can be found using the search facility at http://www01.bhphotovideo.com/

halo surrounding bright objects such as Jupiter and the Moon, and appears in my ShortTube 80 at ×50. At ×114 there is no mistaking the false color that emanates from bright objects. In addition to being aesthetically displeasing, chromatic aberration has two undesirable optical effects, namely the reduction of light grasp and contrast. False color may be annoying (it bothers some observers more than others) but it can be reduced by viewing an object directly on axis. But, disregarding chromatic aberration, just what is the high-power limit of the ShortTube 80? The usual rule of thumb is that the maximum useful magnification is about ×2 per mm, or ×50 to ×60 per inch of aperture under good atmospheric steadiness, known as "good seeing". This means that the ShortTube 80 should reach its magnification limit at about ×160.

However, small telescopes are affected less by poor seeing than large ones and under ideal conditions can be used at much higher magnifications per unit aperture for certain applications, such as resolving binary stars. Refracting telescopes are often better optically than other designs owing to their unobstructed design and other factors, and can sometimes be used at as much as ×4 per mm. Some observers have successfully utilized magnifications above ×200 with their ShortTubes. On fairly steady nights I have been able to observe the binary stars $\varepsilon$ Boötis and $\zeta$ Ursae Majoris at ×250 using stacked Barlow lenses and an 8mm eyepiece. I have also viewed the Moon at the same magnification with better results than I had expected.

The approximately 43mm diameter aperture mask works effectively in reducing chromatic aberration by increasing the ShortTube 80's focal ratio to $f/9.3$, but this comes at the cost of resolution, which is primarily a function of aperture during visual use, and light gathering capability. In my experience views of the Moon are acceptable with the mask in place, but it robs Jupiter of too much light to be useful. The use of a yellow color filter such as a Wratten #8 will also reduce false color.

# Performance Enhancements

## Eyepieces

What are reasonable alternatives to the original eyepieces? I strongly recommend replacing the stock

Kellner oculars with Plössls or orthoscopics. What brands should one consider? Tele Vue Plössls are considered to be among the best available, but the five-lens element Masuyama design Celestron Ultimas are very close in quality and cost less. The similar but more expensive Orion Ultrascopics would also be fine choices. Meade Series 4000 Super Plössls are another possibility. At more modest prices there are a variety of brands to choose from, including Adorama Pro-Optic Plössls, Celestron Plössls, Meade Series 3000 Plössls, Omcon Super Plössls, Orion Sirius Plössls, and University Optic Abbe orthoscopics.

The ShortTube really comes to life when high quality wide-field eyepieces are employed. Premium eyepiece lines such as the Tele Vue Panoptics and Naglers, the Meade Series 4000 Super Wide Angles and Ultra Wide Angles, the Orion Lanthanum Superwides, and the Pentax SMC XLs are certainly worthy of consideration, but are far from inexpensive.

Since a fully dark-adapted human pupil dilates to a maximum of about 7 mm, an eyepiece producing an exit pupil[16] greater than that figure will effectively reduce the aperture of a telescope and decrease the brightness of the view. When using a refractor an observer can "waste" aperture to increase the size of the field of view without the detrimental effects that occur in reflectors and catadioptrics because of the central obstructions created by their secondary mirrors. However, because 80 mm is not much aperture to begin with, I suggest using an eyepiece that will not exceed a 7mm exit pupil.

A 35mm eyepiece will produce a minimum magnification of ×11 in the ShortTube, but since this eyepiece has a field stop no larger than that of a 32mm Plössl, no additional true field of view will result. The maximum usable magnification is usually at an exit pupil of about 0.5 mm, which for the ShortTube 80 means a 2.5mm eyepiece (or a longer focal length eyepiece "Barlowed" to 2.5 mm) and ×160.

I use the following four eyepieces the most often with my ShortTube 80: A 30mm Celestron Ultima, a 19mm Tele Vue Panoptic, an 8mm Tele Vue Radian,

---

[16] The exit pupil is the diameter of the beam of light that converges from the eye lens of an eyepiece, and is equal to the focal length of the eyepiece divided by the focal ratio of the telescope.

and a 7mm Tele Vue Nagler. All but the first of these oculars are each more expensive than the telescope is! Although it is certainly not necessary to buy such expensive wide-field eyepieces, they will help to bring out the best the ShortTube has to offer. Those of you who own more than one telescope, as I do, may have some of these oculars in your eyepiece kit anyway.

For scanning and as a finder eyepiece I use the 30mm Ultima. This is one of the best buys in 1.25-inch eyepieces, and gives me a 3.8° true field of view and a 6mm exit pupil. Also useful for these purposes is the wonderful 19mm Panoptic, which excels at ×21 with a true field of view of 3.2° and a 3.8mm exit pupil. The 19mm Panoptic is my workhorse eyepiece since it produces a great combination of magnification and field of view.

For "high" power I employ two eyepieces. The superb 8mm Radian results in ×50, a true field of view of 1.2°, and an exit pupil of 1.6 mm. The 7mm Nagler produces ×57, a 1.4° true field of view, and a 1.4mm exit pupil. Despite the Nagler's larger true field of view and slightly higher magnification, I prefer the Radian because of its higher contrast and generous 20 mm of eye relief. To see the entire 82° apparent field of the 7mm Nagler I have to press my eye tightly against the eyecup.

I also use a ×2 Ultima Barlow lens and a ×2.5 Tele Vue Powermate to increase the magnification of these eyepieces. Since a ×2 Barlow doubles the focal ratio of the ShortTube, it does help to reduce some of the ShortTube's optical shortcomings. A ×2 Barlow can be used "ahead" of a diagonal to produce approximately ×3 magnification. Unfortunately, both "shorty" Barlow lenses such as the Celestron Ultima and traditional Barlow lenses require the Orion #5123 extension tube ($19.95) to come to focus when used before the diagonal. Similarly, an eyepiece cannot be used "straight through" (i.e., without a diagonal) unless an extension tube is in place.

## Diagonal

The supplied 45° prism diagonal has a diameter of considerably less than 27 mm, the maximum field stop for a 32mm ocular, and thus vignettes the apparent field of view of longer focal length eyepieces. Since it is a roof prism, a so-called optical discontinuity that is inherent in that design can be seen as a dark line bisecting the Airy disk during a star test. This aberration is not visible in daytime use.

One can improve the performance of a ShortTube 80 quite easily by adding a high-quality 90° mirror star diagonal such as the ones sold by Intes, Lumicon ($99.50), and Tele Vue ($75.00). This is perhaps the single most effective way to boost the capabilities of the telescope. Less expensive mirror diagonals are available from Orion (#8778, $55.00), Meade, and other companies. A 90° prism star diagonal does have some advantages over a cheap mirror diagonal.[17] However, according to the noted refractor designer Thomas Back, a prism diagonal can add overcorrected color and spherical aberration at the ShortTube 80's fast focal ratio of $f/5$. However, one should be wary of shoddy, poorly collimated mirror diagonals! A 90° prism diagonal has some advantages over a cheap mirror diagonal, and is a worthwhile alternative.

## Modifications

The more adventurous among you may want to consider making certain modifications that will noticeably improvement your refractor but will also undoubtedly void the telescope's one-year warranty, so proceed at your own risk!

The first fix, and by far the easiest, is to correct for pinched optics. Some ShortTube 80's are shipped with the objective lens tightened excessively. To check for this problem simply perform a star test, using a bright star and high magnification. The star's Airy disk should be circular and not change shape as you rack in and out of focus. A triangular Airy disk means that the plastic retaining ring is too tight. Simply unscrew the ring a little and repeat the star test to see if the problem has been alleviated; if not, slacken the ring some more.

A number of ShortTube owners have removed the objective lens from the lens cell and blackened the edges with a marker pen. This entails completely unscrewing the retaining ring, inverting the optical

---

[17] http://www.egroups.com/files/80f5/diagonals.txt. To access this and other files in the 80f5 egroups community, you need to subscribe to 80f5. It costs nothing; just send a blank email to 80f5-subscribe@egroups.com

tube, and carefully catching the objective lens in a clean, soft cloth. Once the objective lens has been removed, the screws that lie within the optical tube can be painted with mat-black paint.

One of the focuser assembly screws protrudes into the light cone and causes a diffraction spike that is noticeable when viewing bright objects. An easy solution to this problem is to file down the screw a little and to follow through with a dab of mat-black paint. Be sure to remove any metal filings from the optical tube before reassembly. Make sure you reinsert the lenses in the proper order and position.

More complicated "mods" include proper collimation (i.e., optical alignment) of the ShortTube's optics, and changing the spacing and rotation of the objective lens elements to reduce astigmatism and other aberrations. Two amateur astronomers have made available at the 80f5 egroups Website extensive instructions for doing so. Bob Betha's has posted ShortTube "tune up" tips,[18] and Dave Randell has posted his tips for improving 80mm $f/5$ refractors, including a procedure for collimation.[19]

## Accessories

Since one of the ShortTube 80's attributes is the ability to fit large nebulae such as the North America Nebula (NGC 7000) into the field of view in their entirety, I strongly recommend the purchase of a 32mm narrowband nebula filter such as the Orion UltraBlock ($99.95) or Lumicon UHC ($99.50). Such filters will be quite beneficial to the deep-sky enthusiast.

If you plan to use Wratten color filters for planetary observing, I suggest the lighter colors such as the Wratten #21, Wratten #23A, and the Wratten # 80A because of the ShortTube's modest light grasp. While I personally do not find a Moon filter necessary, some people will want one. Orion offers one for $15.95. By all means consider a full-aperture solar filter such as the Orion #7706 ($59.95). One could also make a solar filter using the excellent BAADER Planetarium AstroSolar safety film.[20]

---

[18] http://www.egroups.com/files/80f5/Tuneuptipsfor80F5.txt
[19] http://www.egroups.com/message/80f5/13?&start=1
[20] http://www.astro-physics.com/products/accessories/solar_acc/astrosolar.htm

# Use in the Field

My primary reason for purchasing an Orion ShortTube 80 was to have a portable telescope suitable for viewing the 1998 Caribbean total solar eclipse. I was able to obtain a used ShortTube 80 in good condition and at a good price via the Internet. Despite my many years as an amateur astronomer, this was to be my first real refracting telescope.

The 25mm and 10mm Kellner eyepieces that Orion includes with the telescope were of marginal quality and were immediately replaced by eyepieces from my collection. The 25mm "wide-angle" Kellner seems to lack a field stop, which results in an apparent field of view that is larger than the 40° that is typical for a Kellner. (In fact, the apparent field of the 25mm eyepiece is larger than the 50° apparent field of view of my 26mm Tele Vue Plössl.) The price of this enlarged apparent field is noticeable edge of field astigmatism. Stars at the edge of the field of the 25mm are anything but pinpoints.

The telescope star-tested very well, producing nicely circular Airy disks. Diffraction disks were purple when the focuser was racked inside of focus and yellow outside of focus.

After buying a generic 90° star diagonal I was more than ready to chase the Moon's shadow. I elected to view the solar eclipse from a cruise ship, the MS *Veendam*. On clear nights I had the opportunity to observe many of the best deep-sky objects of the southern hemisphere. I was armed with the ShortTube 80 refractor and a 10 × 50 binocular, 10.5mm (×36) and 26mm (×15) Tele Vue Plössl eyepieces, my ×2 Celestron Ultima Barlow lens, and a 32mm Orion UltraBlock filter.

The ShortTube 80 acquitted itself quite nicely on board ship. Several of the nights were very windy and the sea was choppy, so I limited the magnification to ×15, but on other occasions higher magnifications of ×30 and ×36 could be used to good effect. I used ×72 on the Moon briefly during the occultation of Aldebaran on March 4, 1998, but the result was almost comical as the Moon repeatedly swung in and out of the field of view. There is, after all, a limit to what modern cruise-ship stabilizers can do!

At night the crew cooperated by keeping most of the lights off, and it was really quite dark on the upper forward deck when one stayed in the shadows. M41 was an easy naked-eye object. After we hit the latitudes of the trade winds, observing from the upper deck was out of the question and, of course, when the ship headed northward after the eclipse the aft navigation deck was the place to be. As a matter of fact, my first ever view of the Tarantula Nebula (NGC 2070) and the Large Magellanic Cloud was from this location.

In addition to numerous winter, spring, and summer northern hemisphere deep-sky objects I logged quite a few southern objects, some of which I had never observed before. Before I mention those I should say that portions of the Rosette Nebula (NGC 2237) were visible in my ShortTube 80 when the Ultrablock filter was in place. The Sword of Orion was positioned high in the heavens, and was simply striking telescopically.

I had no trouble star-hopping from object to object using the supplied 6 × 30 finderscope. Among the many southern deep-sky objects I observed with the ShortTube 80 were NGC 2477 and NGC 2451 (a wonderful pair of open clusters in southern Puppis that are remarkably similar to M46 and M47 in the northern section of the same constellation), Collinder 135, NGC 2546, IC 2391, IC 2602, NGC 2516, NGC 2070 and the Large Magellanic Cloud, NGC 2808, NGC 3114, the Eta Carinae Nebula (NGC 3372), the extremely rich open cluster NGC 3532, the Jewel Box (NGC 4755), the Coalsack, Centaurus A (NGC 5128), and Omega Centauri (NGC 5139). Alpha Centauri was easily resolved at ×30, as was another fine southern binary star, $\gamma$ Velorum. The Pipe Nebula (Barnard 59, Barnard 65–67), a dark nebula in Ophiuchus, looked just fantastic through my ShortTube 80. This was by far my best view ever of this fascinating absorption nebula.

The eclipse began in the late morning of February 26, 1998. I was positioned forward on the *Veendam*'s promenade deck; my ShortTube 80 and 35mm SLR camera stood ready on separate tripods. I had decided to view the eclipse through the ShortTube 80 and to take a few photographs through a 200mm telephoto lens. When totality finally arrived I beheld the Sun's almost living crown in awe. The solar corona and polar brushes shimmered and danced in my eyepiece. The corona was utterly fascinating, and a fiery solar prominence was visible at the top of the field of view. As the eclipse progressed, three additional prominences appeared.

**Figure 4.4.** The February 26, 1998 total solar eclipse (after third contact), Pentax K1000.

After what seemed to be only a few seconds I believe that I caught the return of Baily's beads. Then the second diamond ring burst forth majestically as I pulled my head away from the eyepiece and triggered my camera. Third contact – totality was over! Needless to say, those 3 minutes and 36 seconds were the quickest of my life, but I thoroughly enough every second I spent at the eyepiece of the ShortTube 80. As the eclipse progressed past totality I did take some prime focus solar photographs through the telescope (see Fig. 4.4).

Another occasion on which the ShortTube 80 was put to good use was at the 1998 Stellafane convention, an event that is held annually at Breezy Hill in Springfield, Vermont. Under dark skies that were illuminated only by the occasional meteor and satellite pass I observed M27, M31, M32, M110, M33, Stock 2 (the Muscleman Cluster), NGC 869 and NGC 884 (the Double Cluster, see Fig. 4.5), and many other deep-sky objects. Once again, the ShortTube excelled as a rich-field scope. While attending the 1999 Delmarva StarGaze V at the Tuckahoe State Park in Maryland, my ShortTube 80, 26mm Plössl, and Orion UltraBlock nebula filter gave my fellow observers and me a very good view, one that compared favorably to that of an Edmund Scientific 108mm $f/4.2$ Astroscan 2001 Wide Field Newtonian, of the expansive North America Nebula (NGC 7000).

**Figure 4.5.** The Double Cluster in Perseus (NGC 869 and 884): a 2-minute piggy-backed exposure on Kodak Gold Max 400, Pentax K1000, September 11, 1999.

In September of 1999 I had the opportunity to observe from atop the 1483 m (4863 ft) summit of Spruce Knob, the highest mountain in West Virginia. I put my 320mm (12.5-inch) *f*/4.8 Starsplitter Dob to very good use that night, but I didn't neglect my trusty ShortTube. In addition to cruising the riches of the Milky Way from Sagittarius through Cygnus and beyond, I captured a number of deep-sky objects photographically with the

**Figure 4.6.** The first-quarter Moon, photographed on May 22, 1999 with a Pentax K1000 on Fuji Super G Plus 400 film.

**Figure 4.7.** The Orion Nebula (M42 and M43): a 2-minute piggybacked exposure on Kodak Gold Max 400 film, Pentax K1000, September 11, 1999.

ShortTube piggybacked on my friend Sandy Goodstein's Celestron Ultima 8 Schmidt–Cassegrain.

During a Messier marathon that I undertook on the night of March 31, 2000, I easily logged 45 Messier objects using my ShortTube 80 under the very dark skies of Cherry Springs State Park in Potter County, Pennsylvania. The ShortTube's portability came in very handy for capturing those Messier objects that I couldn't access from where my 320mm Dob was positioned. And the ShortTube's wide field of view allowed me to locate many of those objects very quickly, an important factor in Messier marathoning. I was pleasantly surprised to find that I was able to discern the faint open cluster NGC 2158 while I was in the neighborhood of M35. From the same location on May 6, 2000, I detected the eastern (NGC 6992) and western (NGC 6960) segments of the Veil Nebula using my

Orion UltraBlock nebula filter and 30mm Ultima eyepiece.

## Alternative Choices

For the complete novice. the Celestron Firstscope 80 EQ Wide Angle may actually be an even better buy than the Orion ShortTube 80. This Celestron comes with a black finish, an "astronomically correct" 90° diagonal, a 25mm SMA eyepiece, a 6 × 30 finderscope, and, most importantly, a lightweight but reportedly serviceable German equatorial mount and a wooden tripod. The whole package weighs 5.9 kg (13 lb) and has been sold for as little as $228.00.

The Tele Vue 70mm *f*/6.8 Ranger ($695.00 with a star diagonal) and Pronto ($945.00 with a star diagonal) are also worth considering. These scopes have identical optics, but the more expensive Pronto has a rack-and-pinion focuser and accepts 50mm eyepieces. A 20mm Tele Vue Plössl is included with each telescope. The longer focal ratio and use of ED (extra-low dispersion) glass reduces chromatic aberration to a considerable degree.

An 80mm *f*/6 refractor has recently become available from Stellarvue.[21] This $279.95 telescope is known as the Stellarvue 80mm F-6 and comes with 9mm and 25mm Plössls, a 6 × 30 finder, and a 45° or 90° diagonal. As a result of its longer focal length (480mm), chromatic aberration is less of a problem than it is with the ShortTube 80, and higher powers should be more readily usable.

Apogee offers an 80mm ED *f*/6.25 refractor with a 50mm Crayford focuser, 45° prism diagonal, a 25mm Plössl, and a carrying case for $699.00.

Orion Telescope and Binocular Center offers a big brother to the ShortTube, the ShortTube 90mm (3.5-inch) Rich-Field Refractor ($329.00). This *f*/5.5 telescope is reportedly being manufactured in Taiwan and has an optical tube that is 46 cm (18 inches) long and weighs 2.3 kg (5 lb). The ShortTube 90 should have about 27 percent more light-gathering power than that of its sibling. However, reports on the sci.astro.amateur Internet newsgroup indicate that the original version of

---

[21] http://www.stellarvue.com/80mmf6.html

the ShortTube 90 did not have a light grasp superior to the ShortTube 80. Apparently the problem was an overly long drawtube which inadvertently acted as a baffle, a shortcoming that the manufacturer has since remedied.

Discovery Telescopes sells a very similar-looking 90mm *f*/5.6 refractor, the System 90 EQ, for $379.00. For this price you get two Plössl eyepieces (10 and 26 mm), a ×2 Barlow lens, a 45° diagonal, two tripods, an equatorial mount, and a carrying case.[22]

Another of the ShortTube's competitors is the Edmund Scientific AstroScan 2001.[23] This 108mm *f*/4.2 Newtonian comes in a variety of packages ranging in price from $299.00 to $499.00. The AstroScan provides marginally better light grasp than the ShortTube 80, but is even less suitable for lunar and planetary observing.

Some giant binoculars have apertures equal to or even larger than that of the ShortTube 80. Binocular astronomy is a distinctly different experience (two eyes are definitely better than one), and I am a strong advocate of it. Nevertheless, since purchasing my ShortTube I find myself using my Celestron 20 × 80s with decreasing frequency. The ShortTube 80 is more versatile (offering variable magnification and photographic capabilities), more portable, and easier to use than the 20 × 80s with tripod-mounted binocular guider.

## Conclusion

I heartily recommend the 80mm *f*/5 Orion ShortTube to beginners who are seeking an uncomplicated and inexpensive first telescope, to more experienced observers desiring a "quick-look" scope, and to anyone yearning for rich-field views of the Milky Way. The large field of view possible with the ShortTube 80 effectively bridges the gap between binoculars and the vast majority of telescopes. Although it certainly does not have the optical or build quality of, say, a Tele Vue Pronto, the ShortTube 80 is a genuine bargain when

---

[22] http://www.discovery-telescopes.com/TELESCOPES/90rft Rrctr/90RFT.html

[23] http://www.edmundscientific.com/Products/ListProducts. cfm?catid=13

price and image brightness are taken into consideration. Its ease of use, low cost, and surprisingly good performance make it a true "star" in the modern telescope firmament.

# Chapter 5

# Visual Observation of Deep-sky Objects with Small Telescopes

Jay Reynolds Freeman

There is a myth that small telescopes are useless for deep-sky visual observing. If that were not so, people who wanted to view such targets with such equipment would simply get a book on deep-sky observing, and begin. Yet many amateur astronomers think that's impossible, and don't even try. Or perhaps they think it requires a special magic to see faint diffuse objects with tiny telescopes, and they will never find anyone who can give them the necessary charms and incantations.

But it isn't so. Deep-sky work with a small telescope is done the same way as deep-sky work with any other telescope, using the same techniques, and for the same pleasures and purposes. It produces the same kinds of results. There is nothing special about it, except that the telescopes are small. A chapter about deep-sky observing appears in this book for completeness, and to dispel the myth – not to reveal arcane knowledge that applies only to these lesser apertures.

That's not to say that small telescopes don't make the task more difficult. If there is one great truth of deep-sky observing, it is that *aperture wins*, and wins big. As aperture increases, more light enters the telescope, fainter objects become visible, and details show up in objects barely seen in tinier instruments. Fortunately, there are two other, compensating truths. First, you don't need large aperture to do deep-sky observing. There are plenty of interesting objects to look at, even with telescopes as small as 50 mm (2 inches). Second, you can develop observing skills just as well with a small telescope as with a large one. The deep-sky abilities you acquire thereby will be with you throughout your observing career, to help you get the most from any telescope. You can learn them on an inexpensive, compact instrument that is easy to store, transport, set up, and use. Let's start by talking about some of those instruments.

## What Telescope?

For deep-sky observing, the more aperture you have, the better. You will see more sights if your equipment crowds the large end of the aperture range we call "small": for example, a 115mm (4.5-inch) reflector will outperform a 60mm (2.4-inch) refractor. Telescope type is less important. For a given aperture, a refractor will probably deliver more light to the eye than will a reflector or a catadioptric system, because the light lost in passing through lenses, even uncoated ones, is generally less than is lost in several reflections from metal-coated surfaces, and because refractors have no central obstruction. Yet for a given aperture and with given

**Figure 5.1.** Four small telescopes which the author has used for some two thousand deep-sky observations. Back row, on simple altazimuth mountings: 70mm (2.75-inch) f/8 Vixen refractor, 55mm (2.2-inch) f/8 Vixen refractor, and 94mm (3.7-inch) f/7 Brandon refractor. Front: 150mm (6-inch) f/4.7 Newtonian reflector, normally used with no mounting.

accessories, refractors are generally more expensive than other types, and small beginner's refractors are often more bulky, too.

Many inexpensive small telescopes are shoddily constructed. Avoid them. An instrument that looks like a toy probably will work like a toy. Beware of things that are going to wiggle, break, fall off, or are not resistant to wear and tear.

Deep-sky observing requires a solid, vibration-free mounting. The eyepiece should be at a comfortable height and location. You should be able to move the telescope in small increments while you are finding and tracking objects. Mechanically operated slow motions, or perhaps an electrically powered sidereal drive, are desirable for this purpose, yet deep-sky work with small apertures usually requires lower magnifications than with bigger ones, so the apparent drift rate of celestial objects is smaller. Hence the lack of fancy slow motions is not as much of a disadvantage for small telescopes as it is for their larger siblings. Newtonians mounted as popularized by California amateur John Dobson are excellent, and though this style of construction has been used for very large amateur telescopes, Dobsonians as small as 110 mm are commercially available, and are easy to build.

The telescope should be equipped with a sight or finder that will show stars as faint as the ones you will initially use to orient it, in the conditions in which you will be observing. The finder need not show everything you wish to find, for if you can see enough nearby stars to establish the position of an object, you can set the cross-hair on it even when it is too faint to see.

Many people find using a non-magnifying (or ×1) reflex finder, such as the Telrad, particularly straightforward, but if your skies are less than dark, your naked eye may show too few stars for one of these to be of much use. For a magnifying finder, try one with no prism or mirror in the optical path, so that you look into it in the same direction that it points. Practice using it with both eyes open: Let your brain combine the image of the cross-hair, from the finder, with the wide field of view seen by the other, unassisted eye. Use the image of the cross-hair, thus apparently superimposed on the unmagnified sky, to judge which way to move the telescope so as to bring objects into the finder's field. Then use only your finder eye to line them up exactly.

Small telescopes sometimes come with little eyepiece holders – a diameter of 24.5 mm (0.965 inch) is common. There is nothing intrinsically wrong with small-diameter

**Figure 5.2.** The 70mm (2.75-inch) f/8 Vixen refractor with ßuorite doublet objective, mounted on a better-grade altazimuth mounting with slow motions about each axis. A high-end small refractor such as this one will show over a thousand galaxies, star clusters, and nebulae, from a suitable dark site.

eyepieces, but deep-sky observers often use relatively low magnifications, and at normal focal ratios, narrow eyepiece barrels unduly restrict their fields of view. The larger, 1.25-inch (31.8mm) size is much better. Remember also that big eyepiece lenses won't help much if their field is confined by telescope parts, so make sure that any large-diameter eyepiece holder is not merely an adapter at the end of a smaller focus tube.

Simple telescopes sometimes come with just a simple sight for a finder, or with just a couple of protrusions from the tube to squint along. These will do in a pinch – I have occasionally located objects in a Newtonian by using the knobs at opposite ends of the focuser shaft in this way – but if your telescope is so equipped, put something better on your shopping list. Half of an old binocular, or a retired child's toy telescope, would likely be a substantial improvement.

A deep-sky telescope should be particularly well baffled against stray light. There is an easy test: Take your telescope outside by day, point it into the sky well away from the Sun, remove dust cap and eyepiece, and look into the focus tube. You should see a sky-colored circle of light coming in through the objective, perhaps partly blocked by secondary mirror and support. All else should be pitch black. Even closed-tube telescopes, such as refractors, Schmidt–Cassegrains and Maksutovs, may have inadequate internal baffles, allowing stray light to reach the eyepiece. Newtonians with no cap at the bottom will let light sneak past the primary and up the tube. Fortunately, many of these problems are easy to fix. You might rubber-band a paper sack over the bottom of a Newtonian, or extend a dew cap even further skyward, as an external baffle.

Another myth is that fast telescopes – with low $f$-numbers – are better than slow ones – with high $f$-numbers – for deep-sky work. That might be true for photography, for the longer focal length of a slow telescope spreads out the light on the emulsion, but what counts with visual observation is the intensity of the light that falls on the observer's retina. That is determined by aperture and magnification, not focal ratio. A slow telescope requires a longer focal length eyepiece to achieve a given magnification, than does a faster one of the same aperture, but the availability of long focal length eyepieces makes focal ratios of $f/10$ or more entirely practical for deep-sky work. Because of the cramping effect of eyepiece barrel diameter, however, a fast telescope will usually show a wider low-

**Figure 5.3.** The substantially modified 94mm (3.7-inch) f/7 Brandon refractor with Christen triplet objective, mounted on a sturdy altazimuth mounting with slow motions about each axis. A lacy garter and handkerchief help keep a loose dust-cap in place.

magnification field of view than will a slow one, which is an advantage.

# Accessories and Auxiliary Equipment

The dividing line between telescope and accessory is not sharp, since different manufacturers have different notions of what should be built in. Thus I have already mentioned mounts, finders, slow motions, and sidereal drives. There are some more:

## Eyepieces

Other books in this series (e.g., Chris Kitchin, *Telescopes and Techniques*; Stephen Tonkin, *AstroFAQs*) discuss the details, nomenclature, and formulas that apply to eyepieces (or oculars, as they are also called), so I shall confine my remarks to specifics for deep-sky work, such as choice of magnification, and image quality at the edge of the field. As you consider eyepiece purchases, bear in mind that you may some day use them with larger and more expensive telescopes. Thus if your budget permits, it is not necessarily ridiculous to think about buying expensive eyepieces even if your present telescope is inexpensive and small.

Every telescope should have an eyepiece that provides the widest field of view permitted by the diameter of the focus tube, simply to make it easy to find things. That eyepiece can do double-duty for deep-sky work, since many deep-sky objects have such a low surface brightness that they are best seen at low magnification. For these views I suggest an eyepiece with an exit pupil of 4 or 5 mm, which is to say a magnification equal to the aperture of the telescope in mm divided by four or five.

If your telescope has a slow *f*-number, you may have trouble finding an eyepiece with a focal length long enough to provide such a large exit pupil – an *f*/15 refractor would require an eyepiece with a focal length of 60 to 75 mm. Furthermore, the view through such an eyepiece, pared down in diameter to fit a 24.5mm focuser, may be so painfully small that you would rather not use it. My advice is to grit your teeth and put up with a long focal length eyepiece, if you can find one. With faster telescopic $f$-numbers, it is easier to find suitable low-magnification eyepieces. Plössls of 32 and 40 mm focal length are common. They work well at $f/8$ to $f/10$, particularly with a 1.25-inch focuser.

Focal ratios much below $f/8$ begin to cause eyepieces problems: Many common eyepiece designs show deterioration of image quality at the edge of the field at $f/5$. I occasionally use a 20mm Erfle as a low-magnification eyepiece with an $f/5$ telescope, and consider it satisfactory, though star images at the field edges are far from perfect. Much better eyepieces are available for fast focal ratios – the first was the groundbreaking Tele Vue Nagler – but their high cost and considerable weight and bulk are what you were seeking to avoid when you chose a small telescope in the first place. Additionally, many of them require a 2-inch (50.8mm) focuser, particularly at longer focal lengths.

I use a 1.5mm exit pupil for deep-sky work about as often as a 4mm or 5mm one. With such an eyepiece a telescope will have a magnification of about two-thirds its aperture in mm. That magnification will sound high to many, and seems a long jump up from a 4mm exit pupil. Yet if I could have only two eyepieces for deep-sky work, those would be the ones, and they would do for most of my observing.

A 1.5mm exit pupil is particularly well suited to compact deep-sky objects with relatively high surface brightness. The nuclear regions of many galaxies are good examples, as are many unresolved globular and open clusters. That magnification also begins to make available a fair proportion of the telescope's resolving power, so it is well suited to observing many of those star clusters that the telescope can resolve. It is also less bothered by light pollution than are lower magnifications: Background light is spread out, so that more light pollution is required before the background sky becomes luminous enough to reduce contrast. I live in a residential suburb between the great cities of San Francisco and San Jose, California, yet when I observe objects for which a 1.5mm exit pupil is appropriate, I can see them nearly as well from my backyard as from a much darker site.

There are plenty of choices for eyepieces that provide a 1.5mm exit pupil. Users of any focal ratio from $f/4$ through $f/15$ should be able to find an orthoscopic or Plössl with the right focal length. At the longer focal ratios in this range, a less expensive eyepiece type, such as a Kellner, might do. It is also possible that a ×3 Barlow lens, plus your wide-field eyepiece, would be satisfactory, but not all such combinations work well with all telescopes. Try before you buy.

If you have a higher budget for eyepieces, your next deep-sky choice may well be one that gives a magnification of twice your telescope aperture in mm. That eyepiece will give you the best chance of resolving globular clusters into their constituent stars, and may show detail in high surface brightness nebulae that you never suspected was there. Many such nebulae are little-known planetaries, but do take a high-magnification look at the Orion Nebula. Remember, though, that at these magnifications even small telescopes are often limited by seeing, so be prepared to wait for the best views.

Since you will also likely use a high-magnification eyepiece for difficult detail on the Moon and planets, you should not compromise its quality, and should avoid giving any extra lenses, such as Barlows, a chance to degrade its images. My personal choices are high-end orthoscopics, such as those by Zeiss or Pentax, or the Plössl-like Brandons. Orthoscopics from several other manufacturers are almost as good, and much less expensive. There is much competition in the high-quality eyepiece market, so things may change before you read my words.

Good zoom eyepieces may have a place in the deep-sky observers' kit. Models introduced in the late 1990s by Vixen and Tele Vue provide a handy focal length range of 8 to 24 mm, and a reasonable apparent field of view. With such an eyepiece you can make fine adjustments of magnification to maximize the visibility of what you are observing. Changing the focal length by as little as 2 mm often makes a great difference to the detectability of a faint object. Thus a zoom eyepiece substitutes for half a dozen conventional eyepieces. Yet I am not convinced that zoom eyepieces work well with all telescopes, particularly those with fast f-numbers. Again, try before you buy.

## Setting Circles

Setting circles, both mechanical and electronic, provide a straightforward way of locating celestial objects. If the telescope is well built and doesn't bend when you are using it, and if the circles are precisely made and accurately aligned, all will be well. Yet I learned to find deep-sky objects by star-hopping – navigating by referring to star charts as I look through the eyepiece – on telescopes that did not have setting circles. I am proficient enough in star-hopping that I am rarely tempted to use circles when I have them, so I cannot tell you that setting circles are necessary for deep-sky work. Use them if you like.

# Computer-controlled Telescopes

Small telescopes have become available that have altazimuth mounts with electrical slow motions on both axes and can thus track stars across the sky under computer control. These instruments tend to be smaller, less massive, and more expensive than similar units with equatorial mounts. They generally work satisfactorily, and their price has been dropping. Get one if you wish.

More sophisticated computer controls feature a database of interesting objects, and will point the telescope toward whatever you select. If you are not fluent with less automated means of finding things, or if you are trying to locate objects using a small finder in a light-polluted sky, these systems may save you enormous amounts of time. For an experienced star-hopper under a dark sky the speed advantage is less, but I certainly cannot keep up with a Meade LX200 or a Celestron NexStar as it slews rapidly to a new target. Yet my personal observing style is to scrutinize each object for many minutes once I have found it, so an increase in finding speed is not much of an advantage. Even if I could go from one object to the next in no time at all, it wouldn't shorten my observing sessions significantly. Furthermore, this kind of computer control is still pretty expensive. So I haven't bought one of these telescopes.

On the other hand, if I were following an observing program in which I spent only a small amount of time at each target, computer-controlled finding would help a great deal. Examples of such programs include patrolling galaxies for supernovae, or – getting away from extended objects for a moment – observation of variable stars.

# Filters

There are two different approaches to preparing filters to reduce the effect of light pollution. One kind of filter is designed to block light from certain sources of artificial light, and let pass all the rest. The most important kind of light to eliminate is from sodium-vapor and mercury-vapor streetlights. Since these emit a limited range of wavelengths, the range that remains unblocked is rather wide. Thus these filters are called broadband, meaning that they allow a broad band of wavelengths to pass through.

Broadband filters are useful only for light pollution of the kind they block. Where I observe, some neighboring cities have cooperated with the needs of the nearby Lick Observatory by installing street and outdoor lights easily blocked by filters like these. When I point a telescope toward such a location and install a broadband filter, permanent outdoor lights in public places become very dim, but there is still plenty of incandescent light from houses and automobiles. Most celestial objects emit little at the wavelengths of streetlights, so their light is not as much diminished as the light pollution. Hence they become easier to see. Yet if I try the same experiment in a place where most light pollution comes from sources not preferentially blocked by the broadband filter, that filter is nearly useless.

Other filters pass only specific wavelengths of light from particular kinds of celestial object, and block the rest. They are therefore called narrowband filters. They come with several different bandpasses (the range of wavelengths they allow to pass through), corresponding to prominent emission lines in the spectra of emission nebulae associated with star formation, like the Orion and Lagoon Nebulae, or in planetary nebulae like the Ring and Dumbbell Nebulae. When I look at almost any artificial source of light with a narrowband filter, it all but vanishes. How wonderful it is to see dark sky all the way down to the horizon! Yet most celestial targets vanish, too. Narrow-band filters are useful only for objects whose light they pass.

A fair summary is to say that broad-band filters work for some sources of light pollution, and when they do work they will help you see all kinds of celestial object. Narrowband filters work for all sources of light pollution, but they only help with a few kinds of object. What filters you will find useful depends on what sources of light pollution you face, and what kinds of object you like to look at.

There is a specialized use of narrowband filters: Consider a planetary or emission nebula so small you cannot distinguish it from a star, or perhaps the seeing is so poor that you gave up trying. You maybe able to make the distinction by passing a narrowband filter back and forth between your eye and the eyepiece. Stars in the field will have their brightness much diminished by the filter, but most planetary and emission nebulae will not. This technique is called "blinking," even though it is not the nebula that appears to blink, but everything else.

I have an old broadband filter, and a newer, narrowband one. I use the narrowband filter regularly. I use the broadband filter far less often, because when I am observing in sky so bad that it such a filter would help appreciably, I generally choose not to observe deep-sky objects at all. Yet that is my personal choice, and you may choose to do differently.

# Dealing with Dew

Small telescopes have an enormous advantage over big ones for fighting dew: When one of mine dews up, I just carry it into the house or the car for a few minutes till it dries off. I can't do that conveniently with things that weigh more than I do. The only specialized anti-dew equipment I regularly use with small telescopes is a conventional dewcap.

When you take a telescope from a cool place to a warm one, cover the optics first. Let them warm up before you expose them to warm air. If you don't, the cold surfaces may condense out even more dew, so your optics become drippier than you might wish, and take longer to dry. That can happen not only going from the cold outdoors into a warm house, but also going from an air-conditioned dwelling into a muggy summer night.

Finders can dew up, and many have no dewcaps. Improvise one from cardboard and a rubber band, or keep the finder cover on when you are not using it, or simply put a mitten or hat over the finder. Eyepieces also dew. If your pockets are clean or your eyepieces are dirty, you might simply leave unused eyepieces in a pocket to keep them warm. Otherwise, put each eyepiece in its box or container, and put that in your pocket. If you work out of your car when observing, keep the eyepieces in the vehicle when they are not in use.

## Charts, Atlases, and Catalogs

For deep-sky work I use three kinds. First, I keep a small, simple planisphere handy so I can quickly find out what parts of sky will be well placed for viewing on particular dates and times. Second, I use an atlas of naked-eye stars, which has right ascension and declination coordinate lines printed on it, as an orientation aid for those occasions when a page in my big atlas has no star on it whose location I recognize. The naked-eye atlas tells me where in the sky to point my telescope, so that the process of star-hopping to an object may begin. My naked-eye atlas happens to be an old *Norton's Star Atlas*, but many others would do as well.

Third, I use the most comprehensive star atlas I can find and carry, with large numbers of stars and deep-sky objects plotted, all on a generous scale. As I write these words, my choice is Sinnott and Perryman's *Millennium Star Atlas* (Sky Publishing), a three-volume set that costs US$ 250, and outweighs several of the telescopes with which I have used it. Its stellar limiting magnitude of approximately 11 means there are nearly always several of its stars in any field of view I choose, not only when I am using a finder, but also when I am using my main telescopes, at least at low magnifications. With this atlas it is thus very easy to find objects by star-hopping.

I have used several earlier atlases with stellar magnitude limits as high as 9.5. None has enough stars for me always to rely on having some in the field of my main telescopes. Thus I consider the substantial increase in weight, bulk, and cost of the *Millennium Star Atlas* to be well worth it.

On the other hand, the *Millennium Star Atlas* is not perfect – I wish it had better coverage of *New General Catalogue* (NGC) and *Index Catalogue* (IC) objects – and it is heavy and expensive. An alternative criterion for selecting an atlas might be whether it contains at least the stars you can see with your finder. In order of magnitude limit, from faint to bright, suitable atlases include Tirion, Rappaport, and Lovi's *Uranometria_2000.0* (Willmann-Bell), Tirion and Sinnott's *Sky Atlas 2000.0* (Sky Publishing), and the charts in Pasachoff's *Stars and Planets* (Houghton Mifflin).

Many observers use a small personal computer with one or more planetarium programs installed, in the field, instead of atlases. I do not, for a reason that has nothing to do with astronomy: I work in the computer industry, and I like my hobbies not to remind me of my job. Many experienced deep-sky observers rely on nothing else, so I am convinced that computers are a suitable substitute for charts, but I cannot advise on their use.

Your choice of catalogs and other text sources will reflect your equipment, experience, and preferences. Ones I keep handy for deep-sky work, in my order of utility, include the latest copy of the Royal Astronomical Society of Canada's annual *Observer's Handbook*, the three volumes of Burnham's venerable but thorough *Celestial Handbook* (Dover), volume 2 of Hirshfeld and Sinnott's *Sky Catalogue 2000.0* (Cambridge and Sky Publishing), and Sinnott's *NGC 2000.0* (Cambridge and Sky Publishing). There are many other useful books, both specialized and not. Books are among those things that will serve you just as well with large telescopes as with small ones.

## Miscellaneous Aids

A good binocular is a useful deep-sky instrument in its own right, but even a poor one has a particular role to play as a telescope accessory, as a finder for the finder. In light-polluted sky you may not be able to see enough stars to know where to point your finder, much less your telescope. A cheap binocular, on its cord about your neck, will likely let you see all the stars your naked eye would reveal at a darker location, and get you going.

A red flashlight (torch) will save your night vision when you need illumination to consult charts or move about. You can improvise one with a regular flashlight and red nail polish, red transparent plastic, or a red marker pen, but I prefer the kind that have red light-emitting diodes as a light source, with an adjustment for the intensity of the beam.

Some people have difficulty keeping one eye open and the other shut by muscular control, or find it tiring to do. You can always put a hand over the eye you are not using, but you may prefer an eye patch of the sort sold for costumes or medical use, to keep stray light from becoming a distraction.

Creature comforts will improve your visual acuity. Stay warm, and remember that it is easier to keep from getting cold in the first place than to get warm again once you are chilled. I wear many layers of clothes for cold-weather observing, of which a hat is the most important. I use throwaway catalytic warmers in gloves and boots on cold nights (but don't put them where you can't dump them quickly if they get too hot). Sometimes a thermos flask of a favorite hot beverage seems all but vital.

A chair, stool, or short ladder may make observing much easier, particularly if its lack means that you will have to hold your body in an awkward position for a long time, in order to keep your eye at the eyepiece. There are plenty of specialized observing chairs with vertical height adjustments, and I have one. Yet I recommend you first see whether something around your home will do.

## Observing Locations

Of course I cannot give you a list of local locations suited to deep-sky work, but I can tell you how to judge such places, and suggest some hints for finding them. The main requirement is darkness, and since most of you probably live in cities, you will probably have to travel to reach a suitable site. If you have no vehicle you will have to seek others of like mind, perhaps in an astronomy club, and share transport. Do not despair of finding sites: I live in a large and densely populated metropolitan area, yet within a 90-minute drive, I know of several sites where conditions are regularly good and occasionally excellent. Even closer to a city the view may be good in certain directions, such as out over the ocean.

Go up if you can. The higher the site, the less air there will be between you and what you want to see, and the more likely you are to be above low cloud or fog, particularly near the coast where such phenomena are common. Sometimes hills and mountains have clouds of their own, so be aware of local weather conditions, perhaps by using images from meteorological satellites that are available in near real time on the Internet.

Safety and ease of access are important. If you are lucky you may find private land where you can observe in comfort and security. Yet most deep-sky observers of my acquaintance set up in parking lots or adjacent areas, on public lands maintained by governments as parks or open spaces, or perhaps on grounds belonging to universities or other private institutions. We generally find the administrations of such areas to be sympathetic and helpful, though occasionally they have never heard of amateur astronomy and need to be educated about who we are and what we do. It often helps to offer to hold an occasional advertised open observing session for the general public.

# Technique

The ability to see faint objects with a telescope does not come without effort, nor does it necessarily come quickly. There are many techniques to learn. Some are rather mechanical, and easy to explain; once you have learned them, you just need to remember to use them. Others are knacks that seem to develop with practice and experience – at any rate, I have not been able to figure out how to teach them. Here is my personal list of techniques and skills, in order of importance, with the most important things first:

1. *Patience.* It can take a long time to see everything in a field of view, even if you know exactly what to look for and where to look. Many things that affect vision, like seeing, transparency, and eye motions, are not under voluntary control, and can vary quickly. It may take a while before everything adds up beneficially.

2. *Persistence.* Sky, telescope, and eyes vary from night to night.

3. *Dark adaptation.* The adaptation of your eyes to dim light is effected by biochemical changes that may take hours to complete. Even when you have been in darkness for that long, it may help to close your eyes completely for a few minutes before a difficult observation.

4. *Averted vision.* The part of your eye that sees detail best is the least sensitive to dim light. Look a little to the side of what you are trying to see. Don't forget to use averted vision even when you can see

something without it: There may be more there than what is obvious.

5. *Stray light avoidance.* Even when your eyes are dark-adapted, nearby lights make it hard to see faint things. Avoid them. Try eye patches, and eye cups for eyepieces. I once viewed the Sculptor Dwarf Galaxy with my jacket pulled up over my head, with my binocular sticking out. I expect I looked as silly as I felt, but I did see the galaxy.

6. *Changing magnification.* I have already given some hints about what magnifications to use, but in any case the only way to be sure you are using your best eyepiece for the job at hand, is to try them all.

7. *Focusing critically.* Especially at high magnification, you need precise focus to see all the detail. In poor seeing it is hard to focus precisely, but it is worth it.

8. *Moving the telescope.* The eye seems to detect motion, or changing brightness levels, more easily than static images. Jiggle the telescope – move it back and forth. Perhaps you can see an object only while it is moving, or only as the motion stops. Perhaps you can detect the edge of a large object when the intensity of the "background" changes as you move the field. Don't forget to use averted vision while you are doing so.

9. *Not moving the telescope.* The eye seems to be able to add up photons over many seconds. If you can hold it still for long enough, faint things may appear. Don't forget averted vision.

10. *Respiratory and circulatory health.* If you smoke, take a break before and during observing – carbon monoxide interferes with the ability of the blood to transport oxygen. You will probably have a longer observing career, too.

11. *Hyperventilation.* Don't faint, but a series of deep breaths, or at least a deliberate effort not to hold your breath, will put more oxygen into your blood-stream which may improve your ability to detect faint objects.

# Specific Observing Programs

If you are looking for something specific to do, here are a few ideas to get you going.

# The Messier Objects

At least for northern-hemisphere observers, the classic introductory deep-sky program is to observe all 109 objects in the famous catalog compiled originally by French comet-hunter Charles Messier. That list is reprinted in most observing guides, including several I have already mentioned. The catalog is not comprehensive – there are things in the sky that Messier could have seen, but didn't – yet it provides a generous selection of different kinds of deep-sky object, and includes many that are very easy.

I have seen all of the Messier objects with an inexpensive 50mm refractor, but it took pretty good conditions to do so. I have also seen them all with 7 × 50 and 10 × 50 binoculars, though these lesser magnifications made it more difficult to distinguish some of the objects from stars. With a 70mm or 80mm refractor or binocular the Messier list is a cinch, and larger apertures reveal ever more detail. I have made it a habit to go through the Messier catalog with every telescope or binocular I get to use for an extended period. By the time I am done, I am very familiar with the operation of the instrument, and I have learned a lot about how it performs, compared with other equipment I have used to make Messier surveys.

Because of the rather lopsided distribution of Messier objects in the sky, it is possible for a northern-hemisphere observer to see nearly all of them on a March night when the Moon is close to new. This possibility has encouraged manic observers to stay up till dawn on such occasions to see how many they can find. If you are inclined to try such a Messier marathon, have fun – but don't wake me.

As a final Messier challenge, how many of these objects can you see with the naked eye?

# Other Catalog Lists

The annual *Observer's Handbook* of the Royal Astronomical Society of Canada usually has other lists of objects suitable for projects. These include lists of the brightest or best objects of each of several types, a good list of non-Messier showpiece objects, and a few lists of things that are extraordinarily difficult. Once you have figured out what you like to look at, pick one and go to it. Don't expect to find all of the difficult objects with a small telescope.

More comprehensive observer's guides and catalogs have other lists that are organized in useful ways. Burnham's *Celestial Handbook* groups deep-sky objects by constellation. Volume 2 of Sky *Catalogue 2000.0* groups them by type of object. I doubt you can see everything in these books with a small telescope, but their organized presentation will help you plan an observing program to suit yourself.

## The Herschel 400 Lists

In the United States, the Astronomical League has made two selections of 400 objects each from the 2500 or so deep-sky objects discovered by Sir William Herschel. The first Herschel 400 list is a lot of fun. Its objects vary widely in difficulty, from naked-eye past visual magnitude 12, though some published tabulations of Herschel 400 brightnesses use photographic magnitudes, which are systematically too faint. I believe I have seen all of the first Herschel 400 in an exquisite small refractor – a 55mm Vixen fluorite – though it took excellent conditions and every shred of observing skill I possessed to detect the toughest ones. Nevertheless, if your skill, equipment, and conditions leave you bored with Messier objects, try these.

The second list, the so-called Herschel 400 "II" list, is a little confusing since a handful of its objects are "clusters" which modern astronomers do not recognize as true physical associations of stars. The task with these is to try to figure out what asterism or pattern of stars Herschel was looking at. The rest of the objects are of the same general nature as in the first Herschel 400 list.

A final word: I hope you enjoy deep-sky observing with small instruments as much as I do, and that the skills you learn thereby will be valuable even if you move on in your observing career to enormous telescopes. Good luck!

# Part II

# **Reflectors**

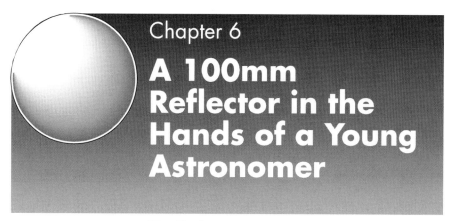

# Chapter 6

# A 100mm Reflector in the Hands of a Young Astronomer

Tim Tonkin

I made the telescope shown in Fig. 6.1, with the help of my dad, when I was ten years old. It is a 100mm (4-inch) $f/5.4$ Newtonian on a simple altazimuth mount. The mirrors are mounted in a square plywood

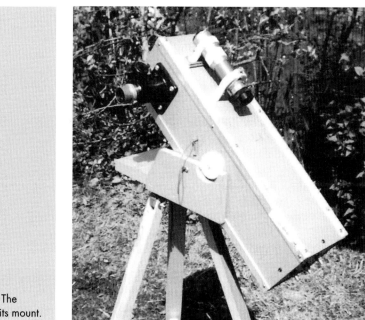

**Figure 6.1.** The telescope on its mount.

tube that has altitude bearings, cut from a plastic chopping-board, on the sides. These bearings sit on blocks of Teflon in the semicircular cut-outs in the mount. These bearings are very smooth, so I use half a clothes-peg as a wedge to adjust the friction. The mount rides on plastic and Teflon bearings on a tripod plate from another telescope. The friction is adjusted by tightening the bolt which holds them together. The tripod legs are lengths of $50 \times 25$mm ($2 \times 1$-inch) softwood. They have central wooden braces which are hinged to the legs and are connected together with an M6 carriage bolt. This tripod is much sturdier than the one which originally came with the tripod plate.

My eyepiece is an old periscope eyepiece and gives a magnification of about ×20. I sometimes borrow my dad's 12.5mm orthoscopic if I want more magnification. The 1.25-inch (31.8mm) rack-and-pinion focuser is the only part of the telescope we bought.

I made the $8 \times 35$ finder from bits of broken binocular, with cross-hairs from my sister's head. It is mounted in plastic pipe brackets. It has two alignment screws instead of the usual three. These work at right angles to each other. and the finder is held against them by rubber bands. This makes it much easier to align because it is easier to think in two directions at 90°, than in three at 120°, and because I don't need to loosen one screw when I tighten another.

To set up the tripod, I extend the legs and secure the braces with their bolt, being careful not to drop the wing nut – finding it in the dark in grass is very difficult. Then I put the telescope on top so that the altitude trunnions rest on the Teflon bearings in the semicircles in the mount. The open end of the telescope extends over the mount.

The next thing is to align the finder, which I usually do using a distant television aerial. When the finder is aligned I sight an object along the corner of the tube, then align the telescope more exactly by using the finder. When I have the object centred on the cross-hairs, I increase the friction (i.e. wedge the clothes-peg into place) and focus the eyepiece of the telescope.

With my telescope I have so far seen quite a few things in the northern hemisphere, from my house in Alderholt, England, and from nearby dark sites like the Isle of Purbeck and Badbury Rings. My favourite objects are open clusters, and I have probably spent more time observing the Pleiades than any other object.

**Figure 6.2.** The telescope being used in "laptop" mode.

I have found many of the Messier objects, but still can't see M101 with this little telescope! I have observed the Moon and all five bright planets. Sometimes I can project the Moon onto a piece of paper. I like to see the craters and to watch the way they appear to change as the Sun rises over them. It is a pleasure to watch Venus slowly go through its phases and to plot the changing positions of Jupiter's moons, especially knowing that I am doing the same things that Galileo did four hundred years ago.

One advantage of such a small Newtonian telescope is that I can use it as a "laptop" telescope with a low-power eyepiece (see Fig. 6.2). It takes practice to find objects when I use the telescope like this, but it is quite convenient to be able to take the tube outside without the mount if I just want a quick look at something in the sky.

I often take the telescope to star parties and field meetings. The eyepiece is low enough for children to use my telescope easily, and I think this makes it easier for them to be interested in what they see. When I tell them that I made this telescope they are usually surprised, and I think that makes them realise that astronomy is something children can be fully involved in. After they've looked through my telescope, some of my friends have asked my dad to help them make one of their own .

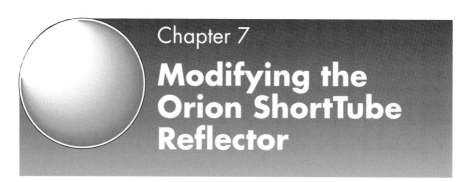

# Chapter 7

# Modifying the Orion ShortTube Reflector

Dwight Elvey

Buying a telescope is always a matter of making choices about price, type, and size. Smaller telescopes like the Orion ShortTube Reflector can have advantages other than just price. It is small and portable, with many functions that make it a desirable first or second telescope. While it is not of top quality, it is comparable to other telescopes in its price range. When you first use it you may be disappointed by some of the "user-unfriendly" features. Here I describe a few simple modifications that can significantly improve its usability.

The ShortTube Reflector telescope is a Newtonian design that has a light equatorial mount and tripod. The optical assembly is made short by using a fast $f/4$ mirror and extending the effective focal length with a built-in Barlow. It also comes with a small ×5 finder and two eyepieces (25 mm and 10 mm). The right ascension (RA) and declination axes have large extended knobs for adjustment.

## Finder

When you first use this telescope you will notice a number of problems. One is that, when you try to adjust the finder, you will find it difficult to get any response from the three adjustment screws. The finder first needs a few modifications. The three screws don't provide a mounting that is adjustable in three dimensions, as

three points always lie in the same plane. We need a
fourth point to complete the mounting. Here is what you
can do. First, remove the finder from the mount. You
will notice that the mount has a slight taper inside that
narrows down on the side away from the adjustment
screws. Take some black plastic electrical tape and wrap
a few layers around the finder's tube (see Fig. 7.1) – just
enough so that it will jam in the taper but is still free to
move some at the wider end. Now we have the necessary
four points: the three adjustment screws, plus the tape,
which provides a pivot at the other end.

Here is a trick that allows you to adjust the finder
during the day, when it is easier to see what you are
doing. The main problem most people have with daylight adjustment is that it is hard to find an object far
enough away not to introduce parallax errors. This is
because any object that is too close will cause the telescope and finder to be angled slightly with respect to
each other when they are both pointed at it. To get
good alignment would normally require the targeted
object to be several miles away. Here is a more practical alignment method: Place a mark in the center of a
piece of paper. Draw a circle around the mark that is
the same diameter as the outside of the telescope tube.
Place the tube vertically on the paper so that it is
aligned with the circle. Now make another mark on the
paper directly below the finder's aperture (see
Fig. 7.2a). I usually make this second mark large
because I will need to see it through the finder from a

**Figure 7.1.** Thin tape
is used to tighten the
finder in its holder.

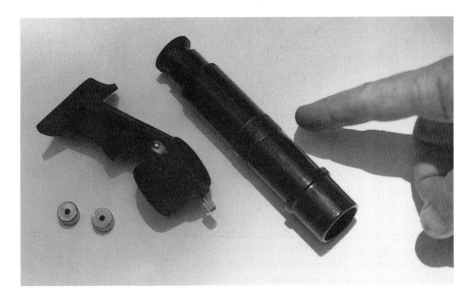

a

b

**Figure 7.2.** How to align the finder in daylight: **a** marking the position of the finder on the paper; **b** mounting the paper on a vertical surface.

distance. Fix the paper to a vertical surface (a wall, say, or a tree) about a hundred meters (300 ft) away, with the marks oriented the same as the telescope tube and finder when the telescope is pointed at the paper (see Fig. 7.2b). Look through the telescope and center the image on the first mark. Now adjust the finder by loosening and tightening the three adjustment screws. Make this adjustment so that the finder points at the second mark. The finder and telescope are now aligned to point at infinity – in our case, planets and stars.

The finder is not of very high quality, and usually has an aperture stop close to the objective lens. It is desirable to get as much light as possible, and sacrifice imaging quality if necessary. I therefore suggest you remove the objective lens assembly and slide the stop about halfway down the tube. This will give more light, though poorer color correction.

## Axis Clamps

That fixes the problem with the finder, but there are still problems. When pointing the telescope at a star, you needs to loosen the RA and declination locking screws. The problem is that, when the screws are tightened down, they transfer some of their rotation to the shaft they are locking onto. The telescope will usually move several degrees, making it hard to find the original object with the fine adjustments. Here is a quick fix: Remove the lock screws and flatten their ends. The manufacturing of threads usually leaves the ends uneven, and with a sharp outer lip that will damage the bearing shafts. You can flatten the ends with a file or by careful use of a grinder. The idea is not to shorten the screw, just make it evenly flat. Making a slight dome at the center of the screw is OK, but try not to make it angled across the screw (see Fig. 7.3). Rolling it between your fingers will show how well you've done. Before putting the lock screws back in place, take a piece of leather shoelace, 3–5 mm ($\frac{3}{16}$ to $\frac{1}{8}$ inch) in

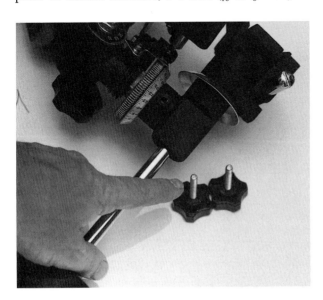

**Figure 7.3.** The ends of the axis-clamp bolts should be slightly domed so that they don't cause the axis to rotate.

length, and put it into the threaded hole. Now when you tighten the lock screw, the leather will lock in place and not transfer the rotation of the lock screw to the shaft. The leather will clamp the shaft and prevent the screw thread from putting dents or grooves into it, and will not shred and clog the bearing. While other materials can be used, I find leather to be the best.

# Increasing the Range of Magnifying Power

You now have a somewhat more usable telescope, but there are still a few more things you can do to make it more enjoyable to use. Many people like to add more eyepieces, which of course cost money. But there is a modification you can make that will extend the usefulness of the eyepieces that come with the telescope. It comes with a built-in Barlow lens, as mentioned above. You may not realize that a Barlow can be used as a variable-power or zoom adjustment. If you get an eyepiece extension tube, it can be like having additional eyepieces (see Fig. 7.4). I use a 114mm ($4\frac{1}{2}$-inch) extension that I bought from Lumicon, but a simple extension can be made from 32mm ($1\frac{1}{4}$-inch) sink drain pipe. Look for one that is tapered slightly, with an external diameter of 32 mm at one end, and an internal diameter of 32 mm at the other end.

Using the extension and the 10mm eyepiece, I get usable magnification up to ×160 from this telescope.

**Figure 7.4.** The extension tube that permits the telescope's built-in Barlow lens to have variable power.

I don't recommend an extension much beyond 100 mm (4 inches) or so; this seems to be about the useful limit of this telescope. When using the extension tube, you will have to move the focus inward to compensate for the increased magnification. I have used this method to watch Io's shadow transit the face of Jupiter.

# Other modifications

Once you have done these things, you'll have quite a useful telescope. I can take it out, make a quick polar alignment, and be observing in just a few minutes. The optics are good enough that on a calm night I can clearly separate such doubles as γ Leonis (one of my favorites). Mostly I have talked about low-cost modification to the telescope, but here are a few more that can significantly improve your viewing, at a cost.

The 10mm eyepiece that came with my telescope did not live up to my expectations. I replaced it with an Orion Ultrascopic that I bought at one of their seconds sales. While this is an expensive eyepiece, the improvement was most noticeable in the sharpness and brightness of the images. I compared the 25mm eyepiece to several of my high-end eyepieces and found it to be acceptable, with no glaring deficiencies.

**Figure 7.5.** The AccuTrack drive connected to the mount.

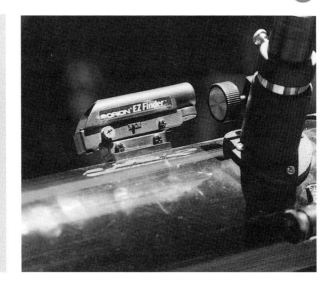

**Figure 7.6.** The EZ Finder mounted on the telescope.

You could also add a clock drive to the RA. This makes waiting for that calm moment when something like Mars is just right much more pleasant. When I bought mine, Orion didn't have a drive specifically for this mount but I was able to adapt the AccuTrack SV to this mount with only slight modifications to the mounting bracket (see Fig. 7.5). When checking for drive compatibility, the main thing to look for in a drive that it is designed to work with a main RA gear with the same number of teeth. Today, Orion sells the EQ-1M drive specifically for the mount that they sell with this telescope. I highly recommend having a clock drive for viewing.

Another enhancement I have added is the EZ Finder (see Fig. 7.6). This is a ×1 pointing device that has a small red LED spot projected at infinity. You simply position the telescope so that the spot coincides with your target, and start viewing. While many prefer the Telrad for this use, I think the Telrad is too bulky for telescopes below about 200 mm (8 inches).

While this isn't the perfect telescope, and I have several others that are larger and of higher quality, I will often pull it out because it is now enjoyable to set up and use. There are a few more things that I'd like to work on, like improving the focuser or adding some damping, but they will have to wait for another day.

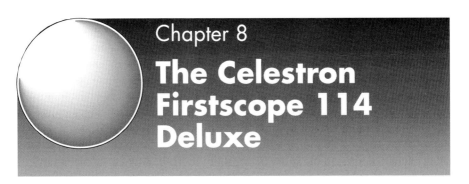

Chapter 8

# The Celestron
# Firstscope 114
# Deluxe

Kevin P. Daly

The universe is filled with innumerable wondrous
sights, many of which are within the grasp of the
Celestron Firstscope 114. I have been fortunate enough
to view many of these for the first time thanks to the
Celestron Firstscope 114 Deluxe (FS 114). This small
telescope has opened up the wonders of the universe to
me and begun what I'm sure will be a lifelong journey
of discovery and awe. In this chapter I endeavor to give
you a brief glimpse of some of the things I've found
possible when using it. In addition I will share the
modifications, accessories, and tricks I've used to get
the most out of this wonderful scope.

The FS 114 Deluxe is a 114mm (4.5-inch) $f/8$ (focal
length 910 mm) Newtonian reflector mounted on
Celestron's CG-3 German equatorial mount. It comes
supplied with 25mm and 10mm SMA (semi-modified
achromat) eyepieces and a 5 × 24 finderscope. I've
found the mount to be more than adequate for visual
use at powers as high as ×266. The one problem I had
in using the telescope in its basic form was the less than
adequate finderscope.

The first thing I did with the mount was to ensure
that all of the nuts, bolts, and so on were tightened. The
first and also the most useful accessory I purchased for
the scope was a Telrad unit power (× 1) finder. This has
aided me immensely in locating the various targets that
I will shortly discuss. I've also added several eyepieces,
a Barlow, and several filters. The eyepieces are a
32mm Plössl, a 6.3mm Plössl, and a 6mm orthoscopic.

I chose the Celestron Ultima Barlow lens to increase my available range of powers for various seeing conditions. The Barlow coupled with the 10mm SMA eyepiece provides ×182, which has proven to be a very useful power when observing the planets. The 6.3mm with the Barlow provides me with ×266 for excellent views of the lunar surface and of some very bright deep-sky objects such as the Orion Nebula (M42). The outstanding contrast of the 6mm orthoscopic at ×150 has provided me with some of my most memorable views of Jupiter. I have found the 1.8° true field of view of the 32mm Plössl wonderful for wide-field views and an excellent aid in locating targets.

# The Moon and the Planets

The first target I sought out was our nearest celestial neighbor. At powers of ×182 and ×266 the detail available is spectacular. Terraced crater walls, craterlets within larger crates, and lava flows in maria all came into crisp view when using this scope. I had never been in such awe of the Moon. The lunar surface provided me with the opportunity to thoroughly learn the workings of the scope, such as aiming, focusing, and polar alignment, and allowed me to get used to the right ascension and declination slow-motion controls.

The FS 114 has proven to be a splendid performer on the planets, routinely outperforming much more expensive instruments. The magnifications that I've found provide me with the best detail have ranged from ×91 to ×182. The planets Mercury and Venus have shown their various phases with startling clarity when viewed at powers from ×133 (using the 6.3mm Plössl) up to ×182 (using the 10mm SMA with the Barlow).

Mars showed me wonderful amounts of detail during its 1999 opposition. Through the 6mm orthoscopic eyepiece (×150), the North Polar Cap, while not obvious, was easily detectable after a few evenings of training my eye. Prominent features such as Hellas and Syrtis Major proved to be easy prey for the scope. I've found that a Wratten #21 orange filter is especially useful in coaxing detail out of elusive Mars.

Jupiter is where the FS 114 begins to really show its planetary capabilities. I had the good fortune on a

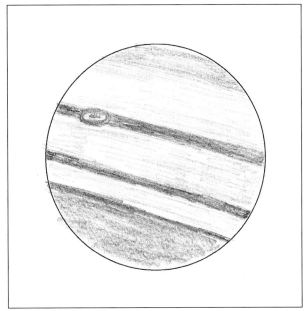

**Figure 8.1.** Jupiter at ×182 (10mm SMA eyepiece with Celestron Ultima Barlow).

summer evening to witness a double transit of the Jovian satellites Io and Ganymede at ×182. The shadows of the satellites appeared very crisp and distinct as I watched them traverse Jupiter's surface. The North and South Equatorial Belts (NEB and SEB) are always quite prominent (see Fig. 8.1), as are all four of the visible satellites even at powers as low as ×91. On nights of particularly good seeing I've counted as many as seven equatorial and temperate belts, as well as being able to see several festoons within the SEB at ×150. The Great Red Spot proved to be one of the toughest details for me to pick out. My first view of this elusive feature was at ×182 with a Wratten #80A light blue filter. Once I knew and understood what I was looking for, the GRS stood out quite prominently with or without the filter at ×150 and up. Jupiter offers a never-ending array of observing opportunities at a myriad of powers with the FS 114.

Saturn is another spectacular target to train the scope on. The Cassini Division is quite evident on nights of fair or better seeing using powers of ×150 and greater. Several belts have also been visible at various times, while observing the "superstar" of the night sky. I haven't found any filters that have proven to be of any benefit while observing Saturn.

While Uranus and Neptune offer little in the way of observing opportunities, they are both easily resolvable into obviously nonstellar disks at ×182. Both show themselves as bluish-green disks at this power. Alas, little Pluto is out of reach of the FS 114, but you still have seven of eight observable planets well within range.

# Nebulae

I've found that the FS 114 provides outstanding views of many planetary and emission nebulae. I always start with the 32mm Plössl to aid me in the initial detection of my target. Once on target I then increase the power until the image begins to either degrade or fade. The brighter nebulae respond very well to higher-power viewing, and I've found quite a wealth of detail in several. I've found narrowband filters such as the Oxygen-III to be of very little use with the scope. They block out so much light that, combined with the small aperture of the scope, they render invisible many nebulae which are otherwise easy.

The Ring Nebula (M57, NGC 6720; see Fig. 8.2) is one of my favorite planetaries to observe with the FS 114. On nights of good seeing the donut shape and dark center of the nebula are quite visible at powers as low as ×91. I've had several evenings on which the Ring

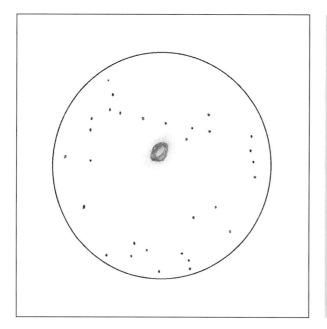

**Figure 8.2.** The Ring Nebula (M57) at ×91 (10mm SMA eyepiece).

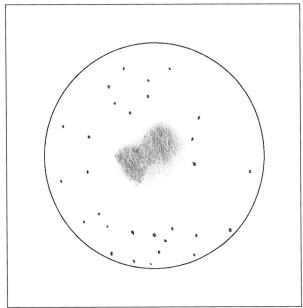

**Figure 8.3.** The Dumbbell Nebula (M27) at ×91.

looked superb at ×182, and even higher on rare occasions.

The Dumbbell Nebula (M27, NGC 6853; see Fig. 8.3) is another wonderful target. At powers of ×91 and greater I've managed to distinguish the dumbbell shape, which is quite a treat.

The Orion Nebula (M42, NGC 1976) is simply spectacular. At the heart of the nebula lies the Trapezium, a group of four stars arranged in a squarish pattern. I've found this detail quite easy to pick out on virtually any night at powers as low as ×91. At low powers the distinctive butterfly shape virtually fills the 1.8° field of view of the 32mm Plössl. This is one nebula where I always seem to find new details at all powers. I've had tremendous views of the filamentary nature of the nebula at ×266.

The Cat's Eye Nebula (NGC 6543) is an excellent target to see color in. At low power in the FS 114 it appears stellar in nature, but at high power the nebula becomes obviously nonstellar with a green tint.

The Lagoon (M8, NGC 6523) and Trifid (M20, NGC 6514) Nebulae in Sagittarius are also outstanding targets. Both contain large open clusters embedded within the nebulosity and are quite easy to view. They are objects I regularly return to with the FS 114.

I've observed dozens of other nebulae with the FS 114 and haven't been disappointed. In particular, the Eagle, Swan, Eskimo, and North America Nebulae are all worthwhile targets.

Although not really nebulae, there are a couple of supernova remnants that show up quite well in the FS 114. A dark site is an absolute must for the Crab and Veil Nebulae. The Veil can prove to be a spectacular sight from very dark skies through the 1.8° field of view of the 32mm eyepiece, filling the view from edge to edge. The Crab Nebula is simply a misty patch in Taurus, but the fact that I'm looking at the remnant of a supernova that exploded less than a thousand years ago is a treat in itself.

# Globular Clusters

While many globular clusters appear as little more than cotton balls, I have found quite a few that resolve into individual stars quite nicely. Increasing the power as high as conditions will permit helps to turn the cotton balls into spectacular countless pinpoints of light. Another technique I've found that helps immensely in resolving globulars is to use averted vision. By looking slightly to the side of my target and allowing the light to fall on the most sensitive part of the eye, I've found that some otherwise unresolvable globulars begin to resolve into individual stars.

The Hercules Cluster (M13, see Fig. 8.4) is one of my favorite targets for the FS 114. With the 6.3mm Plössl at ×133 it begins to show its outermost members around a speckled core. On nights of good seeing at ×182 and with averted vision, I've seen stars right through the core.

Other favorites in this category are M3, M5, and M15 (see Fig. 8.5), as well as most of the Messier globulars in Ophiuchus. All of these resolve quite nicely with the FS 114. I've found that the best technique is to experiment with different powers, and to spend time studying them. There has been many an evening where I've spent an hour or more observing a single globular. The results are well worth the effort.

# Open Clusters

Open clusters have proven to be some of my favorite targets with the FS 114. The 1.8° field of view that the

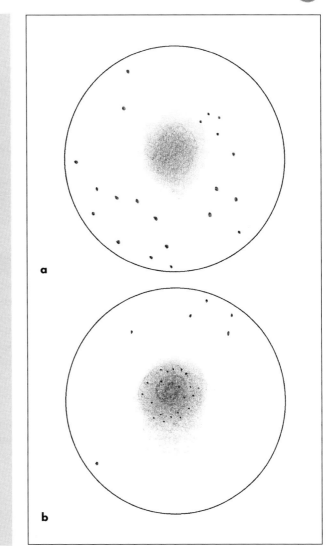

**Figure 8.4.** M13 in Hercules **a** at ×36 (25mm SMA) and **b** at ×91 (10mm SMA).

32mm Plössl eyepiece provides, plus the light-gathering power of the 114mm aperture, is a wonderful combination for this pursuit. This is an area where the FS 114 routinely outperforms telescopes of much larger aperture.

Very large clusters such as the Pleiades (M45) in Taurus and the Beehive (M44) in Cancer each fit quite nicely into the field and provide a stunning view. I've found that low power is sufficient to provide wonderful views of most clusters.

The Double Cluster (NGC 869 and 884, see Fig. 8.6) in Perseus is another favorite. This pair of clusters fits quite nicely in the field of view within a rich back-

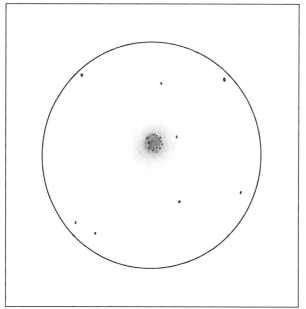

**Figure 8.5.** M15 in Perseus at ×91.

ground of stars. I've compared the view through the FS 114 with those through scopes of much larger aperture, but the Double Cluster seems best framed and presented with the smaller aperture combined with the wider field of view afforded by the FS 114.

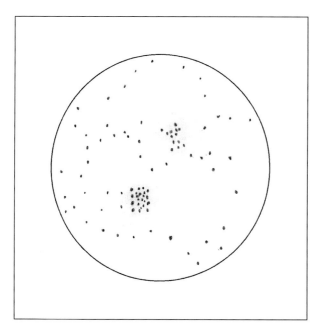

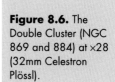
**Figure 8.6.** The Double Cluster (NGC 869 and 884) at ×28 (32mm Celestron Plössl).

I've found that the key to observing open clusters is to experiment with various powers. As with most searches, I always start out at my lowest power (×28). Most clusters will jump out of the background quite well at this power. Some will resolve better at high power, yet offer a much more aesthetically pleasing view at lower powers. Again, experiment with various powers to see which ones provide the best view for the particular cluster you are looking at. There are many targets to pick from in this category, and I've yet to find one that didn't provide a marvelous view.

The "ET Cluster"' (NGC 457) in Cassiopeia is a wonderful target for the FS 114 at star parties. I always enjoy the comments from new observers who see this cluster for the first time. My wife refers to it as the "Pterodactyl." The stick-figure shape stands out quite nicely, and is well framed in the 32mm Plössl eyepiece with the two "eyes" showing quite brightly.

The open clusters NGC 1647 and NGC 1746 in Taurus are two more wonderful targets for the FS 114. Both exhibit a wonderful, compact structure and are easy prey for the scope.

The three clusters in Auriga (M36, M37, and M38) provide a wonderful contrast in the different types and structures of open clusters. All three resolve quite nicely in the FS 114. At low powers (×28 or ×36) they appear as compact hazes of stars, but as you increase the power the clusters begin to reveal themselves in all their glory.

Another worthwhile target is M29 in Cygnus (see Fig. 8.7).

## Galaxies

Galaxies are among the most challenging objects to observe regardless of the aperture of the scope. For most of these most challenging but rewarding objects, their typically low surface brightness, combined with what many of us have to put up with in the way of light pollution, make a trip to a dark site essential. The good news is that under dark skies there are many galaxies within reach of the FS 114. While they don't show a lot of detail in the eyepiece, just knowing that I'm looking at objects millions of light years distant gives me an immense thrill. They do show their various types quite well, ellipticals and spirals standing out quite nicely.

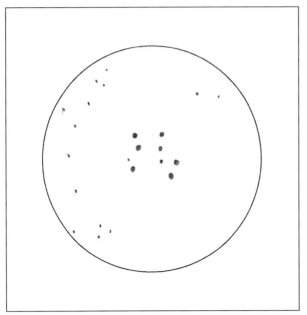

**Figure 8.7.** M29 in Cygnus at ×91.

A favorite pair of mine are M81 and M82 in Ursa Major (see Fig. 8.8). They are framed quite nicely in the 32mm eyepiece at ×28, and readily show the difference between a spiral galaxy (M81) and an irregular galaxy (M82). M82 will even reveal some mottling at high powers (×150 or ×182) under excellent conditions.

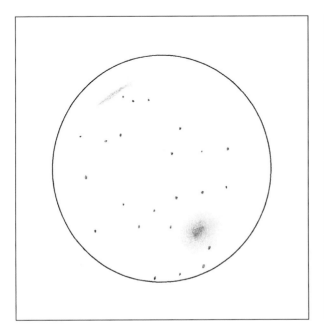

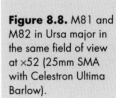

**Figure 8.8.** M81 and M82 in Ursa major in the same field of view at ×52 (25mm SMA with Celestron Ultima Barlow).

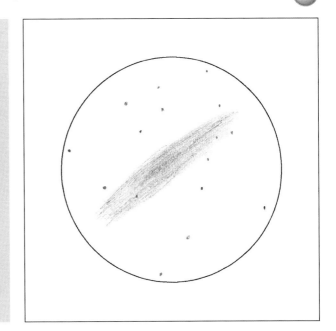

**Figure 8.9.** The Andromeda Galaxy (M31) at ×56 (32mm Plössl with Barlow).

Another galaxy, especially interesting in the wide field of view afforded by the 32mm Plössl, is the Andromeda Galaxy (M31, see Fig. 8.9). This very bright, very large galaxy can typically stretch from one side of the field to the other from a good dark site. Also visible in the same low-power field of view are the largest of Andromeda's companions, M32 and M110.

The Whirlpool Galaxy (M51) and its companion (NGC 5195) are also easily detectable in the 114 mm of aperture. Here is a glimpse of a nice pair of interacting galaxies. A challenge is to observe the bridge between the two galaxies. I've managed to glimpse the bridge using averted vision at ×91 on several occasions.

The Virgo Cluster of galaxies is a real treat with the 32mm Plössl. Scanning this area will reveal dozens of galaxies, many of them visible as groups of six or seven within the same field of view.

Other favorites with the FS 114 are M109 in Ursa Major, M65 and M66 in Leo in the same field of view, and The Sunflower Galaxy (M63) in Ursa Major. These are all quite bright and fairly easy to find with this scope.

Hopefully I've given you an idea of what is possible with a scope such as the Celestron FS 114 Deluxe. Even though I've since graduated to larger scopes, I often find myself grabbing the FS 114, as it is such a joy to use. It will always be a special scope to me, as it hooked me on this wonderful hobby and has provided and continues to provide countless hours of enjoyment.

# Part III

# Catadioptrics

# Chapter 9

# The Mighty ETX

Michael L. Weasner

Meade Instruments Corp. introduced the ETX Astro Telescope in the spring of 1996. I purchased one in September of that year, and soon came to realize just how mighty this little telescope is. A few days later I created the first ETX-only site on the World Wide Web.[1] It has since become the Internet's most popular ETX site. Even though the ETX acts like a much larger instrument (as implied by the photo from my Website, Fig. 9.1), it really is just a 90mm (3.5-inch) telescope.

**Figure 9.1.** The mighty ETX.

---

[1] http://www.weasner.com/etx

There are currently three models of the ETX: the original ETX (now known as the ETX-90RA), the ETX-90EC, and the ETX-125EC. The ETX-90EC was introduced in January 1999 and the ETX-125EC in May 1999. The ETX-90RA and ETX-90EC models have the same overall specifications, the EC model adding motorized slewing in right ascension (RA) and declination, and the option of electronic focusing. The ETX-125EC has the same look and operation as the ETX-90EC but is larger. (The ETX-125EC, and the ETX-60AT and ETX-70AT released in 2000, are not covered here.) Both EC models can accept the Meade Autostar electronic GO-TO computer for automatically locating and tracking some 12,000 astronomical objects and artificial satellites. All models have as standard a motorized RA drive that provides reasonably accurate tracking of celestial objects for visual use. Table 9.1 gives the detailed ETX specifications as supplied by Meade. With specifications like these it is not surprising that the ETX quickly became a success, not only with the casual observer who occasionally views the night sky but also with both the semi-serious and the serious amateur astronomer.

## The Casual Observer

## ETX Astro (ETX-90RA)

The original ETX combined in a small telescope excellent optical quality and a battery-operated drive supplied as standard – not an optional extra. This combination, while available to varying degrees in other telescopes, was not offered for under US$ 500.00 until the ETX was introduced. And even with the $100.00 price increase a few months later, it still offered the best performance for the money. The ETX quickly became one of the most talked-about and popular telescopes for the casual user. For many, the ETX was their first telescope, and the experience of seeing the Moon or the rings of Saturn in such crisp detail and not having to continually move the telescope to keep the object in view was awesome. Even users new to astronomy had success using the ETX. Its small size meant that owners could set it up at a moment's notice to observe or show their friends objects in the night sky. Larger telescopes purchased by casual users tend to

**Table 9.1.** Specifications of the Meade ETX-90RA and ETX-90EC

|  | ETX-90RA | ETX-90EC |
|---|---|---|
| Optical design | Maksutov–Cassegrain | Maksutov–Cassegrain |
| Clear aperture | 90 mm (3.5″) | 90 mm (3.5″) |
| Focal length | 1250 mm (49 inches) | 1250 mm (49 inches) |
| Focal ratio (photographic speed) | f/13.8 | f/13.8 |
| Near focus (approx) | 3.5 m (11.5 ft) | 3.5 m (11.5 ft) |
| Resolving power | 1.3 arc seconds | 1.3 arc seconds |
| Super multi-coatings | standard | standard |
| Limiting visual stellar magnitude | 11.7 | 11.7 |
| Maximum practical visual power | ×325 | ×325 |
| Optical tube dimensions (diameter x length) | 10.4 × 27.9 cm (4.1 × 11 inches) | 10.4 × 27.9 cm (4.1 × 11 inches) |
| Secondary mirror obstruction | 27.9 mm (1.1 inches); 9.6% | 27.9 mm (1.1 inches); 9.6% |
| Telescope mounting | Fork type, double tine | Fork type, double tine |
| RA drive motor | 4.5 V DC | 12 V DC |
| Hemispheres of operation | North and south, switchable | North and south, switchable |
| Slow-motion controls | Manual; RA and dec. | Electric; 4-speed; RA and dec. |
| Bearings | RA and dec., nylon | RA and dec., nylon |
| Tube body | Aluminum | Aluminum |
| Mounting | Reinforced high-impact ABS polymer | Reinforced high-impact ABS polymer |
| Primary mirror | Pyrex glass | Pyrex glass |
| Correcting lens | BK7 optical glass, grade-A | BK7 optical glass, grade-A |
| Telescope dimensions | 38 × 18 × 22 cm (15 × 7 × 9 inches) | 38 × 18 × 22 cm (15 × 7 × 9 inches) |
| Telescope net weight[a] | 4.2 kg (9.2 lb) | 3.5 kg (7.8 lb) |
| Telescope shipping weight | 5.6 kg (12.4 lb) | 5.6 kg (12.4 lb) |
| Price (US dollars) | $595 | $595 |

[a] 90RA including tripod legs; 90EC including electronic controller and batteries.

end up in the closet as the novelty wears off and the inconvenience of moving a large telescope sets in.

So what is this new ETX user experience like? My first impression, as recorded in September 1996, was that the ETX is a beautiful telescope, and is great if you want portability, ease of set-up, and some basic astronomical viewing and photography. I bought some Meade accessories with the telescope. It comes with a 26mm eyepiece (×48); I added a 9.7mm eyepiece (×128), a ×2 Barlow lens, a 45° erecting prism, and a T-adapter for a camera. The telescope and accessories are of high quality. Optically, the telescope has already provided me with some amazing views, even through the brightly lit and hazy Los Angeles sky. Views of the Moon, Jupiter, Saturn, and Venus are very crisp and bright, even at the higher magnifications. Stars and Jupiter's moons are nice points of light. Operationally, the only minor difficulty I experience is with the drive motor. When first turned on, it can take up to a minute to engage fully. Once engaged, it does not always re-engage right away after positioning the telescope on a new object. Once it does engage, tracking is adequate at low power.

After I moved in the summer of 1997 to a new location with darker skies, I noted some additional impressions. My previous location had been nearly at sea level, with lots of heat sources disturbing the air, and so was not ideal for observing. My new location, at an elevation of about 365 m (1190 ft), has dark steady skies that really demonstrate just how optically great the ETX is. Dim stars are crisp pinpoints of light; bright stars have a nice Airy disk with several interference rings visible.

I originally used a very sturdy microwave table on wheels for holding the ETX. It worked great. I just wheeled the scope out the door onto the patio, lined up a compass (adjusted for the local magnetic variation), removed my homemade ETX cover, and *voilà* – I was ready to observe.

On the night of August 22/23, 1997, I spent several hours observing and photographing Jupiter and Saturn. The seeing was incredibly steady. Using a Wratten #82A light blue color filter, cloud bands on Jupiter and Saturn really stood out. I could see the Cassini Division in Saturn's rings clearly, as well as the shadow cast by the rings on Saturn's disk. The optics of the ETX continue to impress me. Figure 9.2 shows two afocal projection images that were among my first attempts at planetary photography with the ETX and a digital camera. (The view through the telescope is much clearer than these digital images might suggest.)

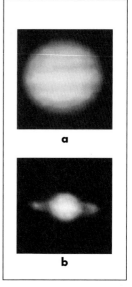

**a**

**b**

**Figure 9.2.** Jupiter and Saturn photographed on August 23, 1997 with a Casio QV-10 digital camera through a Meade 9.7 mm eyepiece with a ×2 Barlow lens. For Jupiter **a** I used two stacked color filters (Wratten #82A, light blue, and #8, light yellow; for Saturn **b** I just used the #82A light blue filter.

# ETX-90EC

In January 1999, Meade surprised the ETX community by announcing a new, improved ETX, designated the ETX-90EC. The following description is from my initial ETX-90EC impressions, recorded shortly after I received one from Meade in early 1999. The ETX-90EC comes with a revised manual, an ETX accessory catalog

(Meade's), and a registration card. The manual is a great improvement on the original ETX manual and includes more details on polar aligning and troubleshooting. The optics and optical tube assembly (OTA) remain the same high quality as with the original model ETX. The drive base and fork mount are totally new but still have a lot of plastic. The ETX-90EC comes with an 8 × 21mm viewfinder, a 26mm Plössl eyepiece, and a new electronic controller (see Fig. 9.3) as standard. The tabletop legs that were standard with the original ETX model are no longer included, but are available separately.

Using the Electronic Controller, you can slew the ETX-90EC in altitude (up/down) and azimuth (left/right) for terrestrial usage. There are also focus in/out buttons on the Controller for use with the optional Meade electronic focuser. You can change from altazimuth to polar mode (northern or southern hemisphere), which engages the RA tracking motor. You then use the controller to slew in RA and declination. Unfortunately the setting is not retained once the power is turned off. To set the tracking mode, you press the MODE and SPEED buttons in a certain sequence, so you have to remember and repeat this sequence for every power-on. You can, however, set a default tracking mode by removing one or two screws from the controller's backplate, as described in the manual.

You need to go through some initial set-up steps, including testing the drive motors to be certain they

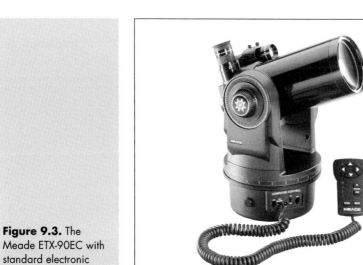

**Figure 9.3.** The Meade ETX-90EC with standard electronic controller.

work properly. I tested them, and all was well. I then took the ETX-90EC outside and got in some observing before the clouds came in. Optically, the views are of the same excellence as with the original model. When you set the controller for polar mode and turn on the power, the ETX accurately tracks smoothly and quietly in RA. There are four slewing speeds for both RA and declination, the highest being for large movements across the sky when you are locating objects manually. In fact, the manual implies that this is the preferred way to move the ETX tube, rather than unlocking the RA and declination locks and manually moving the scope. I was surprised at this, but the speed is actually fast enough to make this practical. The only drawback is that the drives are rather loud at this high speed. The second-fastest speed is for minor movements to position objects accurately in the viewfinder or with a low-power eyepiece. The two slowest speeds allow for precise positioning with higher-power eyepieces or for correcting for a not-quite-accurate polar alignment. Using the electronic controller to slew the ETX will move the scope in a manner more appropriate to terrestrial observing than astronomical observing. Pressing the Left or Right button results in left or right movement of objects in the viewfinder. But since the view through the eyepiece is reversed left-to-right, the buttons seem to work backwards. Once you get used to this, it becomes only a minor nuisance for astronomical observing. At the two slowest speeds both the RA and declination motors sound "hesitant" in that there is no constant volume or frequency to the motor sound. Functionally, both motors slew smoothly, so I guess everything is as it should be.

Manual focusing is the same as with the original model ETX, although it seems to be "tighter" in the new model. When you mount the telescope on the (optional) Meade tripod, hand-induced vibrations (with the 26mm eyepiece) are minimal, but the optional electronic focuser would make a nice addition (just as did the JMI MotoFocus on the original model).

As many users have experienced with the original model ETX (now called the ETX-90RA), views through the ETX-90EC with the supplied 26mm Plössl eyepiece are sharp and thoroughly enjoyable. And just as with the addition of the Microstar dual-axis drive corrector to the original ETX, having electronic control over both the RA and declination axes in the new model makes using the ETX-90EC even more enjoyable.

**Figure 9.4.** The Autostar controller.

If you elect to purchase just the basic ETX-90EC, you will have a fine astronomical or terrestrial telescope. For astronomical viewing you will want a sturdy tripod or the optional tabletop legs for polar-mounting the ETX – that is, unless you add the optional Autostar computer controller ($149.00), which can drive the ETX-90EC (and ETX-125EC, as well as the DS series telescopes) in RA, even in the altazimuth mounting mode.

## Autostar

The Autostar (see Fig. 9.4) is a computerized hand controller providing celestial object selection, automatic slewing of the ETX to that object, and tracking of the object in RA. It includes a database of some 12,000 objects, and allows the user to add more objects such as satellites and comets. It can also be used for automatic slewing to fixed terrestrial objects for users who want to do that. There is a "tour" mode in which the Autostar selects objects currently visible in the night sky, and slews to them and tracks them in turn. When you track an object, the Autostar displays some details about it (e.g., name, location, and magnitude). The Autostar gives you much of the GO-TO capability of the larger LX200 line of Meade computerized telescopes.

The Autostar replaces the standard electronic controller. It is larger than the controller, with more buttons and a display screen, and has nine slewing speeds instead of four. There is even a small red LED on the end that can be turned on and off. The initial set-up is straightforward and quite simple, as long as you follow both the printed instructions and those displayed on the Autostar's screen. This initial set-up takes just a few minutes and, except for entering the date and time at each power-on, you do not need to repeat it, except if, for example, you are going to attach the Autostar to a different ETX, correct errors in the drive training, or change your observing location.

The Autostar's display is easy to read, even in bright sunlight (for the initial set-ups). The letters are white on a black background, which helps a lot. You do not need to use the manual at all times since there are clearly worded prompts on the display and, once you know the basics, it is easy to navigate through the

Autostar menus. The Autostar controller itself is light-weight and easy to hold and use. It would have been nice if a hook had been included so that the controller could be hung on the tripod right side up instead of dangling upside down by the coiled cable. Following the initial (daytime) set-up, you need to do the rest of the alignment at night (when the display is dimmed) and every night that you want to observe, assuming you want to use the GO-TO computer to locate objects for you or even just to track objects in RA.

I found, as did many other users, that there were operational aspects of setting up and using the Autostar that prevented it from working as well as expected. Sometimes you would not see selected objects in the field of view of the supplied 26mm eyepiece. Sometimes the Autostar would report that the selected object was "below the horizon" when in fact it was easily visible in the night sky. On the other hand, many other users reported very positive results in both the polar and altazimuth modes. I experienced some nights of bad results and other nights of pretty good successes. Using the Autostar in the altazimuth mode is likely to give the beginner the best chance of success as it is then less prone to the errors that can occur when attempting to use it in the polar mode. And it should be noted that RA tracking in the altazimuth mode is excellent (for visual purposes).

At the time of my initial Autostar experiences, I felt that, until all users could report positive results "out of the box," adding the Autostar Computer Controller would be a gamble. It could give you the wonderful thrill of seeing many astronomical objects with a minimum of effort, or it might result in total frustration. But you should not hesitate to purchase the standard ETX-90EC. It remains an optically fine astronomical instrument and, with the addition of the motorized controls, will give even the casual amateur astronomer many hours of viewing pleasure for a reasonable initial investment.

I do not know how the Autostar does its calculations, but you should think carefully before either manually selecting alignment stars or accepting the Autostar's selections. Dr. Leon Palmer from Rigel Systems reminded me that the region around the celestial pole is one of the worst places to select an alignment star because the lines of RA converge at higher declinations. This convergence means that a small error in angular position translates into a large error in the RA value. Errors in position are compounded and can even result in an opposite alignment when trying to align on Polaris, which is not precisely at 90°N declination. Dr. Palmer advises using stars that are about 90° (6 hours) apart and near or on the celestial equator for the best results.

Using the #505 connector cable set from Meade and a personal computer, you can update the Autostar software to new versions and add new objects to its database. The cable set includes a 0.9m (3ft), 4-pin to 4-pin cable for use when cloning the software from one Autostar to another; and a 1.8m (6ft), 4-pin to 6-pin cable to connect between the Autostar and a computer to download updates into the Autostar or to control the telescope from the computer (using appropriate software). To connect this cable to the computer's serial port, an adapter is included that mates the 6-pin connector to a DB-9 connector. If you have a PC you likely have everything you need. If you have a Macintosh you will need one more adapter (unless you make your own cable). I used the serial adapter that was included with my Ricoh digital camera. This adapter converts from the DB-9 on the Autostar adapter to the standard Mac mini-8 serial port. USB is not currently supported on either PCs or Macs.

The current Autostar updater is a Windows-only application, although a Java application that will run on Macs, PCs, and Unix workstations has been mentioned on Meade's Autostar update Web page. To run the current updater using a Macintosh you will need either SoftWindows from Insignia or VirtualPC from Connectix. I have successfully used VirtualPC 2.1.3 with Windows 98 on my G3/300 MHz Macintosh. Since the Autostar/computer interface is via the serial port, you will need a USB/serial converter if you plan to use a computer having only a USB port.

I plugged the Autostar cable (using both adapters) into the printer port of my Mac and configured VirtualPC to use the printer port as COM1. Once VirtualPC was set up and the updater installed and running in Windows 98, I turned on the ETX and followed the instructions in the cable set manual. I set the Autostar to "download" and began the process. When the download was complete, the Autostar beeped and began its initialization sequence. I then went to the Autostar Setup/Statistics menu and saw that the Autostar software had been updated.

Meade continue to release new versions of the Autostar software on their Website. They also post current ephemerides for comets, satellites, and asteroids. The current version of the software allows you to obtain, edit, and manipulate the asteroid and comet ephemerides as well as the satellite two-line-element (TLE) files, and obtain, edit, and manipulate Autostar "tours" of the evening sky. Beginning with version 1.2g of the Autostar software, and continuing through version 2.0i (the current version at the time of writing), most ETX-90EC users have reported significantly improved alignments and GO-TO abilities. I have also had considerably more reliable results with this newest version. You will find that objects are now almost always centered in the field of view of the 26mm eyepiece, whether tracking in altazimuth or polar mode. It has also been discovered that the ETX battery level can affect the accuracy of the GO-TO operation. You should replace the batteries if the level is down to around 80 percent, and especially if the level drops even lower while slewing the ETX with the Autostar. It's important that you check the battery level frequently when using the Autostar, or add the optional AC adapter if you find errors occurring. With the revised Autostar software, users who want to experience more of what the night sky has to offer should consider the purchase of the Autostar at the time the ETX-90EC is purchased, or as soon as additional monies become available.

## Traveling with the ETX

One of the reasons I chose the ETX instead of a larger telescope is that I knew I wanted to take it with me on trips. Since I was away from home at the time of the September 1997 lunar occultation of Saturn, I put the ETX to the travel test – and it came through with flying colors. I put the ETX in my Meade soft case. I also packed all my eyepiece cases, papers, and flashlight (torch) into the bag. It was still easy to carry, and easily fit under the airliner seat. The only glitch was when the security staff at Los Angeles Airport wanted to verify that it was a telescope. I do not know what it was the x-ray showed that made them nervous, but a security person asked me to unpack the telescope and remove it from its case. She wanted to be able to see straight through it! I told her that she would not be able to do that and that all she would see was herself. She did not seem to believe me. When I removed the objective lens cover and held the ETX up for her to look down the tube, she laughed and said, "OK!" She also decided she wanted me to open up all the eyepiece cases so that she could see what they contained. She finally agreed that the ETX was not a dangerous piece of contraband and let me repack everything. So, a word of caution: Be prepared to unpack everything if you take the ETX flying. Today there are many cases, both hard and soft, from Meade and other dealers, explicitly for the ETX. You can even get a backpack case from Shutan Camera & Video. I used this case to carry the ETX-90RA when I traveled to Australia in late 1999.

## Why an ETX?

So you are thinking about buying an ETX-90RA or ETX-90EC, but are wondering just what you can see with it and how easy it is to use. First off, forget about all those really neat and beautiful astrophotographs taken with the Hubble Space Telescope, the 200-inch (5m) Hale Telescope, and CCD imagers on 10-inch telescopes. Even if you could look through the eyepiece of those telescopes, what you would see would not match the details and colors that you see in long duration exposures or CCD images. OK, so you will not see swirling, multicolored nebulae or distinct clouds on Mars – but can you take photographs like these? The truth is "not really," although you can accomplish some amazing things with the right equipment and lots of patience. But, if you cannot see all these incredible details, and you cannot photograph them either, why get an ETX?

There are two great qualities of the ETX: It is small (meaning "portable") and optically it is incredible for its size and price. You can see cloud belts on Jupiter and perhaps Saturn (on a good night), the rings of Saturn, the Cassini Division in the rings, some moons of Jupiter and Saturn, phases of Venus, some hints of dark areas on Mars (and the North Polar Cap during a favorable opposition), some galaxies and nebulae (fuzzy though they will appear), double stars (some very beautiful), globular clusters, the Moon, and, with the proper protection, the Sun. If you try, you can actually see a lot with the ETX. But you could see and photograph more details and colors with a larger (much larger, actually) telescope, so why not get a larger telescope?

If you can get a large telescope *and* you are going to use it regularly, great – go for the largest telescope you can afford and use. I stress "use" because many amateur astronomers get excited about having a large telescope, but then tire of the process of moving it outside, setting it up, using it, and moving it back inside. Very quickly, large telescopes can become unused unless the user is really dedicated. The ETX is extremely portable; you will be able to use it on a moment's notice and, with its high optical quality, you will *want* to use it. One other piece of advice for those new to astronomy: it takes some work to really enjoy it. You have to study the skies to know your way around, even if you are planning to use an Autostar to locate objects. You have to read books, get some star charts or star charting software, and learn the language. This is not difficult, nor do you have to be an expert in it. Having the convenience of the ETX actually makes learning the skies more productive. Someday you can move up to a larger scope as your enjoyment of astronomy increases (along with your bank account), but you can still use the ETX on trips or for those spur-of-the-moment observing sessions.

Overall, the ETX provides excellent performance and value for money. While the optical quality is very high, Meade use a lot of plastic in the product to keep costs down. In general I have not found this to be a problem, although a product with the popularity of the ETX is bound to experience some glitches. For the casual amateur astronomer, the ETX is a worthwhile purchase and should provide years of viewing pleasure.

## The Semi-serious Amateur Astronomer

The excellent optical quality and its price earned the ETX many favorable reviews following its release. It quickly became a very popular telescope for all levels of amateur astronomers, and a large third-party after-market arose, offering accessories specifically designed for the ETX. Companies such as Jim's Mobile Inc. (JMI), Scopetronix, Apogee Inc., and Clear Night Products, to name a few, designed excellent add-ons to enhance the ETX usage for the semi-serious amateur astronomer who was willing to spend $50, $100, $150, or even $200 on individual items. I discuss a few of these products next.

There were two notable deficiencies of the original ETX. One, the size and location of the finderscope limited its usability, as did the fact that it was not a right-angle finder. Two, there was no electronic control of the RA and declination axes. Third-party products addressed both of these shortcomings. Apogee and JMI offered similar solutions for the finderscope: A user-installable modification of the ETX finderscope to put its eyepiece at a 90° angle to the ETX tube. For many users this solved the finderscope problem. Other users preferred the "×1" or "red-dot" style of finder from Rigel Systems, Telrad, Scopetronix, and Orion Telescope and Binoculars, or even riflescopes from Daisy.

JMI offered the first electronic axis controller for the ETX in their MotoDec product. This added a motorized drive to the declination axis and a hand controller. Using this simple add-on, you could make minor adjustments in declination while the standard RA drive tracked an object. But there was still no way you could electronically adjust for RA tracking errors, at least not until Scopetronix released their Microstar product. The Microstar modification was more serious in that the user had to replace the original RA drive controller circuit board with the Microstar controller circuit board. This is actually a fairly simple operation. With the Microstar installed, both axes of the ETX could be controlled via a small hand controller. JMI and Microstar offered electronic focuser add-ons that could be controlled from their hand controller. These products addressed another common complaint about the ETX: The image being viewed would bounce around while the user was attempting to focus using the small, standard focus knob.

With the release of the ETX-90EC, Meade made the capabilities of these third-party add-ons standard (dual axis electronic control) and optional (electronic focusing and right-angle finderscope).

Other useful ETX-specific add-ons became available soon after the release of the ETX. Clear Night Products developed the "TeleWrap/Dew Cap," which added a nice astronomical photograph onto the ETX OTA as a functional dew shield. Apogee provided several optical accessories, including adapters for CCD and camera mounting, and a wide-field adapter to increase the viewing area and reduce the magnification of eye-pieces. JMI released a tripod and wedge (for polar-aligning the ETX) and later a "wedgepod" tripod/wedge combination specifically designed for the ETX. JMI also released the first piggyback camera adapter which allowed a 35mm film camera to be mounted on the ETX; the RA drive could track sufficiently well for wide-field, moderately long-duration astronomical photographs using only the camera's optics. When you combine a piggyback camera adapter with the Microstar dual-axis controller and an illuminated reticle guide eyepiece, you can take excellent photographs of the night sky. Two examples are shown here. With just a normal 55mm lens on the camera you can take a photograph such as that in Fig. 9.5, showing the heart of the Milky Way. Going a step further and using a 230mm telephoto lens on the camera (which makes for a rather interesting system, as seen in Fig. 9.6a), you can get nice photos of the more prominent deep-sky objects, such as M42 in Orion (see Fig. 9.6b).

Using the Autostar, astronomy charting software, or astronomical charts, separately or in conjunction with one another, the semi-serious amateur astronomer can explore the heavens, finding new objects almost every night (assuming, of course, that the weather cooperates). Faint nebulae, galaxies, comets, asteroids,

**Figure 9.5.** The Milky Way, f/2.0 55mm lens, 15-minute exposure on ISO 800 film.

**Figure 9.6.**
**a** Piggyback tele-photo/camera on the Meade ETX-90RA, used to photograph **b** M42 in Orion, f/4.5 230mm lens, 10-minute exposure on ISO 800 film.

a

b

Mercury, Uranus, and Neptune can all be seen with the ETX. As the user's experience and dedication grows, so will the range of objects that can be observed, increasing the user's enjoyment. However, the ETX is still only a 90mm (3.5-inch), so it does have its limitations on usable magnification and resolution. But, as noted before, the portability of the ETX can mean a lot, since any telescope is more useful when it is actually used. And adding some astrophotography capabilities will mean that new opportunities and challenges will always be present.

## The Serious Observer

For the serious amateur observer on a limited budget, the ETX has a lot to offer. If you already have a larger telescope, the ETX can make an effective guidescope when mounted on the larger telescope. Some users have attached a CCD imager to the ETX, running the CCD output to a computer, resulting in an automated guidescope that drives the larger telescope for exposures of very long duration.

Other serious amateurs have added more and more to their ETX in the way of accessories and third-party add-ons rather than purchasing a larger, more capable telescope. The extra weight of a lot of accessories and camera/CCD equipment demands a heavy-duty tripod. But one drawback becomes evident very quickly: the drive assembly is not designed to handle much extra weight. This should not surprise you, given the purpose and price of the ETX, but that does not stop users, myself included, from trying. Adding proper counterweights and a replacement drive mechanism will be necessary if you wish to attempt long-duration astrophotography.

The really serious amateur probably already has several telescopes available. But many have still purchased the ETX, partly for its beauty, partly for its portability, and partly for its functionality. For them, the ETX has demonstrated that it really is a "mighty" telescope.

## Photography with the ETX

The original ETX was available in a "spotting scope" model, sold without the drive base. It was basically just the OTA, finderscope, and the 26mm eyepiece. This provided an excellent visual telescope for terrestrial use. You could use an optional erecting prism accessory from Meade to correct the image for terrestrial viewing. While this was a fairly expensive but excellent visual telescope, many users found it worthwhile to use the ETX OTA as a 1250mm telephoto lens on a 35mm camera. If you have the full "astro" model of the original ETX, you could still remove the OTA from the fork mount and attach it to a sturdy tripod for terrestrial use, either for viewing or as a telephoto lens. Meade also has spotting scope versions of the ETX-90EC and ETX-125EC.

Adding a T-adapter allows the ETX to be used for prime focus photography[2] (see Fig. 9.7a), both terrestrial (as a telephoto lens) and celestial. You mount it at the rear of the ETX by removing a screw-on cover from the ETX and attaching the adapter. A separately purchased T-mount ring is required for the specific camera model to be attached. These are inexpensive (about $15) and available from a local camera store. The T-adapter is actually two adapters in one, with one short piece screwed into another short piece. If you use the shorter mount alone, your photos will be less than full-frame. You can do prime focus astrophotography, but you will usually be limited to short exposures (see Fig. 9.7b). The drive does not have the accuracy for long exposures without noticeable trailing on the image. Unless you mount the ETX on a very heavy-duty tripod, movements of the focal-plane mirror and shutter in a 35mm camera will induce vibrations that will ruin any photograph, even short exposures. So it is necessary to use the "hat trick" method of covering the objective end of the ETX, opening the camera shutter, flipping the cover out of the way for the duration of the exposure, re-covering at the end of the exposure, and then releasing the camera shutter.

The basic camera adapter allows you to mount a 35mm camera for eyepiece projection (see Fig. 9.8a). The adapter has a setscrew for holding an eyepiece inside the larger diameter tube. The same T-mount ring used with the T-adapter will also fit the basic camera adapter. You need to remove the adapter from the camera to insert and remove eyepieces. The standard 26mm eyepiece that comes with the ETX is 70 mm ($2\frac{3}{4}$ inches) in length and is 6 mm ($\frac{1}{4}$ inch) too long for normal use in the adapter. It sticks out past the end of the adapter and prevents my Pentax Spotmatic's mirror from flipping up. Not being able to use the

---

[2] Strictly speaking, this is photography at the Cassegrain focus, not the prime focus.

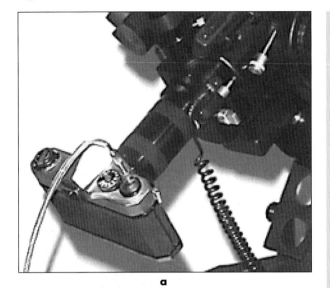

a

b

**Figure 9.7. a** 35mm camera positioned at prime focus with a T-adapter, used to photograph **b** the Moon, prime focus, $\frac{1}{4}$-second exposure on ISO 800 color print film.

26mm eyepiece is not the drawback that it seems since you get nearly the same magnification at the prime focus position. However, you can manually hold the camera reasonably steady against the basic camera adapter for short (less than 1 second) exposures. Shorter eyepieces such as the Meade 9.7mm fit OK, but a Barlow lens is too long to use. The 9.7mm eyepiece is

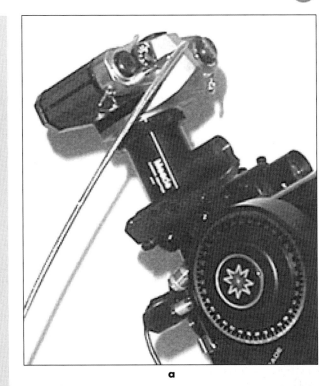

a

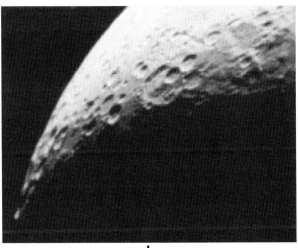

b

**Figure 9.8.**
**a** Camera mounted for eyepiece projection photography using the basic camera adapter, and used to photograph **b** the Moon, 9.7mm eyepiece, 2-second exposure on ISO 800 film

a short eyepiece and so fits way down inside the adapter tube, putting your eye at quite a distance from the eyepiece. The poor eye relief of this eyepiece means that it is a challenge to find objects with the eyepiece inserted into the camera adapter before you attach the camera. And once you insert the camera, you will find it almost impossible to locate faint objects in the camera's viewfinder. You must therefore align the finderscope accurately. However, once you have found your target and focused the image, you can begin to take photographs. Longer exposures will be required if you use eyepiece projection, and you will need to minimize vibrations. But you can get successful photographs this way, as shown in Fig. 9.8b, although it takes patience and a lot of film.

I have already discussed piggyback astrophotography, which is probably the most rewarding photography you can do with the ETX. Some users have attached CCD imagers and even the inexpensive QuickCam video camera to their ETX and have obtained amazing results.

Finally, perhaps the simplest way to take photographs of the Moon, Sun, and brighter planets is to use a digital camera. I have used a Casio QV-10 (my first digital camera) and a Ricoh RDC-4200 (my second one) with the ETX. The Casio actually produced the better results; the Ricoh lens focus and exposure electronics seem to be too sophisticated and

**Figure 9.9.** Lunar eclipse, Casio QV-10 digital camera, 26mm eyepiece (afocal).

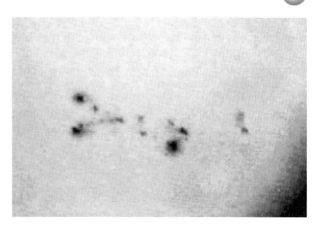

**Figure 9.10.** Sunspot grouping, Casio QV-10 digital camera, 9.7mm eyepiece (afocal), Thousand Oaks Solar II Type 2 Plus filter for used eye, telescope, and camera protection.

so complicate matters. Using a consumer digital camera to do afocal projection photography is straightforward: Focus the object in an eyepiece with your eye and then position the camera lens (focused at infinity) over the eyepiece and take the picture. With luck (and a proper exposure) you can get pictures similar to these shots of the Moon (see Fig. 9.9), the Sun (see Fig. 9.10), Venus, the four brightest moons of Jupiter (see Fig. 9.11), and an occultation of Saturn by the Moon (see Fig. 9.12).

I took these shots by hand-holding the digital camera over the eyepiece, but some users have built devices to hold the camera in place. With the digital camera adapter (DCA) from Scopetronix (see Fig. 9.13) you can do better afocal projection photography with almost any telescope/digital camera combination. Mounting the DCA on your telescope and your camera on the DCA is simple; but, as noted in the instructions, it will take time to determine the proper positioning, given the flexibility (really "adjustability") of the DCA. It is this flexibility that allows the DCA to work with various camera styles, although it does not work with all cameras. When the camera is mounted and aligned you have what is shown in Fig. 9.13 (the camera is a Casio QV10 digital camera).

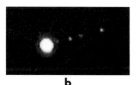

a

b

**Figure 9.11.**
**a** Daytime photograph of Venus, Casio QV-10 digital camera, 9.7mm eyepiece (afocal); **b** Jupiter, four brightest moons all on the right side (visible in the original photo), Casio QV-10 digital camera, 26mm eyepiece (afocal).

With my Ricoh RDC-4200 digital camera I was unable to align the optical axes of the lens and the eyepiece. This camera has a lens that must be rotated 90° into position, significantly shortening the distance from the camera base. However, I was able to take some photos with this camera (with the 26mm eyepiece), thus demonstrating that it was sufficiently well aligned. To take them, I first turned on the camera's self-timer (to allow vibrations to dampen out). I then set the telephoto lens to its maximum (to avoid serious vignetting, or edge shadowing). When the shutter was pressed, the timer began its countdown, the vibrations went away, and then after a few seconds the picture was taken. All the time I could monitor what was happening on the camera's LCD.

You can also use the DCA for terrestrial photography. Although I do not have any interesting land objects near my home, I did take a view of a tree (just to prove it could be done) using the 26mm eyepiece. In this full-frame photo (see Fig. 9.14) you can see the vignetting that appears when using the Ricoh camera (due to the distance of the camera lens from the eyepiece). Unless you can get your camera lens (not just the camera body) close to the eyepiece, you may experience similar vignetting.

The Scopetronix digital camera adapter can be used with most any eyepiece. I coupled it with the Meade 9.7mm eyepiece, and could easily see Venus on the Ricoh's LCD. (Unfortunately I was unable to adjust the Ricoh sufficiently to prevent Venus from being overexposed.) With the large-barrel 40mm eyepiece from Scopetronix I needed to spread the DCA eyepiece clamp wider than normal to make it fit. You can mount the DCA wherever an eyepiece is mounted. Figure 9.15 shows it mounted on the Shutan wide-field adapter with the Scopetronix 40mm eyepiece.

If you have attempted unsuccessfully (like me) to make a digital camera mounting adapter, check out the Scopetronix DCA. Check with Scopetronix to verify that your camera can be optically aligned. If the distance from the optical axis of your camera lens to the base of the camera is between about 40 and 80 mm (1.5 and 3.25 inches), then it should mount and align OK. If

**Figure 9.12.** Occultation of Saturn, Casio QV-10 digital camera, 9.7mm eyepiece (afocal).

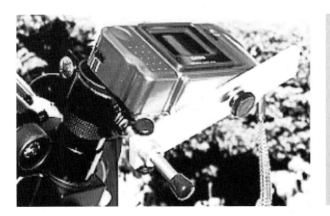

**Figure 9.13.** The Scopetronix digital camera adapter.

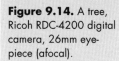
**Figure 9.14.** A tree, Ricoh RDC-4200 digital camera, 26mm eyepiece (afocal).

it does, you will likely get much better results than when hand-holding your camera over the eyepiece. And even though it is limited in capabilities, digital camera photography is so much simpler than traditional film photography. If you have a digital camera, you will eventually want to take photos through your telescope. Once you get frustrated with hand-holding the camera, you will either make your own adapter or purchase the Scopetronix one.

Photography with the ETX is both challenging and rewarding. It is by no means impossible to get reason-

**Figure 9.15.** A loaded "mighty ETX."

able results. With some work (and luck), you can take very nice photos.

# Online ETX Resources

As with all online resources, what is available today can be gone tomorrow, what is good today can be horrible tomorrow, and what is said today can change tomorrow. With these caveats, Table 9.2 lists some useful Websites, a mailing list, and newsgroups that cover the ETX and accessories. There is a continuously growing ETX community online. ETX users worldwide can learn tips (such as using a clothespin on the small focus knob to reduce vibrations during manual focusing), find out how to do astrophotography with the ETX, have questions answered, discover modifications that can be made (e.g., adding a drive-on indicator LED, or tuning up the ETX), and read reviews covering add-ons and accessories. My ETX Website (see Fig. 9.16) is proud to be listed among the best.

**Figure 9.16.**
Weasner's Mighty ETX Site home page.

**Table 9.2.** Internet resources for the ETX

| Resource | URL/name |
|---|---|
| Weasner's Mighty ETX Site | http://www.weasner.com/etx |
| ETX Mailing List | http://etx.listbot.com |
| ETX Web Ring | http://nav.webring.yahoo.com/hub?ring=etx&list |
| General Astronomy Newsgroup | sci.astro.amateur |
| Meade Telescope Newsgroups | alt.telescopes.meade.lx200 |
| | alt.telescopes.meade |
| Meade Advanced Products Users Group (MAPUG) | /http://MAPUG.com |
| Meade Instruments Corporation | http://www.meade.com |
| Scopetronix | http://www.scopetronix.com |
| Shutan Camera & Video | http://www.shutan.com/ |
| Pocono Mountain Optics | http://www.poconoscopes.com/index.html |
| Jim's Mobile Inc. (JMI) | http://www.jimsmobile.com |
| Clear Night Products | http://home.earthlink.net/~barrycnp |
| Oceanside Photo & Telescope | http://www.optcorp.com |
| Rigel Systems | http://www.rigelsys.com |

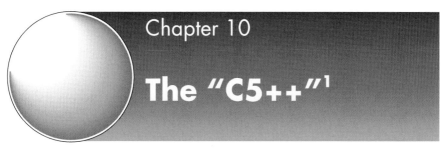

# Chapter 10

# The "C5++"[1]

Stephen Tonkin

## Description

The Celestron C5 optical tube assembly (OTA) has been in production since 1971. Since then, the telescope has come in a number of different varieties, with various mounts and accessory options as standard. These include:

- *Original C5* — Mains-driven double-tine fork, $6 \times 30$ finder.
- *C5 Spotting scope* — Tripod mount, accessory bar and $6 \times 24$ finder.
- *C5 "Classic" scope* — Mains-driven single-tine fork mount, table-top equatorial wedge, accessory bar and $6 \times 30$ finder.
- *C5+* — As the C5 "Classic", but battery driven.
- *G5* — Lightweight German equatorial, tripod and $6 \times 30$ finder.
- *NexStar 5* — Computerised altazimuth mount and unit power reflex finder.

The C5 Classic and C5+ are no longer produced, but are sought-after second-user instruments, considered by many amateur astronomers to be classic portable telescopes.

---

[1] The sobriquet "C5++" was suggested by a correspondent and refers to the high level of "accessorisation" to which I have subjected the original telescope.

**Table 10.1.** Specifications of the C5's optical tube assembly

| | |
|---|---|
| Aperture | 125 mm (5 inches) |
| Focal length[a] | 1250 mm |
| Focal ratio[a] | f/10 at Cassegrain focus |
| Focusing | Movable primary |
| Length | 280 mm (11 inches) |
| Weight | 13 kg (6 lb) |
| Rear cell[b] | 48mm male thread |
| Resolution | 0.9 arc second |

[a] The effective focal length and, therefore, the focal ratio depend on the separation of the primary and secondary mirrors. In a Cassegrain telescope in which focusing is achieved by moving the primary mirror, stated focal lengths and ratios are thus nominal values whose precise values will depend on the position of the focal plane. Used with a star diagonal, my C5 has an effective focal length of about 1290 mm (f/10.3).

[b] The 48mm thread on the rear cell accepts a variety of accessories including a visual back (standard equipment), a T-threaded photographic back, an off-axis guider, and a f/6.3 focal reducer/coma corrector.

The C5 OTA (see Table 10.1) has a deserved reputation for optical and mechanical excellence. One of the pitfalls of focusing by moving the primary mirror in Schmidt–Cassegrains is that, unless the mechanism is very precisely machined, rocking of the mirror during focusing introduces image shift. No such shift is detectable in my C5, and reports from other users suggest that this is the norm for this instrument. In addition, the focusing is exceptionally smooth – another quality which reports suggest is the norm.

The optical quality is excellent. A star-test suggests that there is a tiny amount of undercorrection, but this is sufficiently small for it not to contravene the Rayleigh criterion[2] and prevent the telescope from performing at close to its theoretical limit on a night of steady seeing.

# Mounting

My C5 is the C5+ incarnation, obtained second-hand. The OTA is held by a dovetail joint to the tine of a

---

[2] The Rayleigh criterion stipulates that an optical system whose wavefront error is less than one-quarter of a wavelength will, in use, be indistinguishable from a perfect system.

single-arm fork mount. The fork is itself mounted on the drive base, and has a male $\frac{1}{4}''$-20tpi threaded attachment for counterweights. The fork–drive assembly has manual slow motions in RA and declination and an electric spur-gear drive in RA. The electric drive can be powered by a PP3 (9V) alkaline battery that sits in a recess in the drive base, or by an external 12V supply. I have found the drive to be extremely good. When it is properly polar-aligned, the mount maintains sidereal rate with less than an arcminute of periodic error and with no abrupt jumps. (I have never bothered to quantify this precisely, but it is in the region of 30 arc-seconds). It is also very quiet, unlike the drives of some other commercial Schmidt–Cassegrains – a definite advantage if one does not wish to disturb the neighbours on warm "open window" nights!

The drive base attaches to an equatorial wedge which itself serves as a table-top mount. It has an integral circular bubble level to assist in levelling. Like the fork mount, the wedge is of extremely sturdy metal construction.

The weakest link in the mount is the tripod. I have the Celestron "field tripod" that is dedicated to this telescope. It is adequate for visual work, but not for long-exposure photography or CCD imaging. For these purposes I use it with the legs fully retracted and standing on vibration suppression pads. The accessory tray that attaches to the tripod braces adds more rigidity. This is not ideal, but the arrangement is usable with care.

## Accessories

One of the advantages of having a respected telescope of reasonable age is that there are a number of dedicated accessories available. A disadvantage of the "reasonable age" aspect is that some dedicated accessories are no longer made. I use the following with my C5:

*Reflex Finder* This replaces the original 6 × 30 finder scope. I am not a fan of small finders. and the reflex finder is sufficiently accurate for me to be able to locate my target object in a low-power field.

*Declination Motor* An essential for guided long-exposure photography and imaging.

**Figure 10.1.** Solar observation with the C5+ and Baader solar filter.

*Hand Paddle* This is required to operate the declination motor. It also modifies the rate of the RA motor by up to three times the sidereal rate, and is useful for centring an object as well as for guiding.

*Solar Filter* A home-made filter, using Baader Planetarium Astro-Solar film, fits over the aperture (see Fig. 10.1). This is an excellent-quality filter material and yields very crisp, white solar images.

*Dew Shield* Dew is a curse of Schmidt- and Maksutov-type telescopes, in which the corrector plate behaves as though it were specifically designed to attract dew! A dew shield is therefore essential. The flock-lined plastic "wraparound and Velcro" type is not only ideal as a dew shield, but also packs flat, making transport easy.

*Counterweights* I was unable to obtain the dedicated counterweight set, so I made my own from a length of 25mm (1-inch) diameter mild steel rod and some collars that were designed to secure barbell weights to a bar (see Fig. 10.2). This combination is very versatile and is sufficient to counterbalance any accessories I attach to the telescope. I have made a smaller counterweight which attaches to the accessory bar and balances the telescope about the declination axis. This is much more convenient than sliding the entire telescope along the dovetail when adjustments are needed during an observing session.

**Figure 10.2.** The home-made counterweight bar.

*Piggyback Attachments* These allow a 35mm camera or a CCD camera to be attached to the accessory bar, either directly or via a ball-and-socket joint (see Fig. 10.3).

*Focal Reducer/Coma Corrector* This dedicated attachment reduces the focal ratio to a nominal $f/6.3$ and improves the quality of the outer parts of the field of

**Figure 10.3.** Piggyback attachment in use.

view. It also permits a brighter image of extended objects with my low-power eyepiece, a 32mm Plössl.

*Guidescope* I originally used a 60mm refractor for this purpose, but it is quite heavy and is much longer than the C5 OTA. More recently I have made a more compact one from a 300mm camera lens and a ×2 tele-extender to which I cemented a star diagonal (see Fig. 10.4). It has an aperture of 56 mm, and its focal length of 600 mm gives a magnification of ×67 with my 9mm guiding eyepiece, which is adequate for guiding when the focal reducer is on the main telescope. The addition of a Barlow lens doubles the effective focal length, and therefore also the magnification, of this excellent little guidescope. I made the rings by which it fits to the accessory bar from old metal drainpipe brackets.

*Off-axis Guider* The off-axis guider, also known as a radial guider, inserts a small adjustable prism into the light path and deflects a small amount of the light into a guiding eyepiece. It eliminates any flexure between the main telescope and a guidescope, and guiding automatically compensates for any primary mirror shift. However, only a very small amount of light is deflected,

**Figure 10.4.** C5+ with Starlight Express MX5 CCD camera and small guidescope.

making it an interesting exercise to find a suitable guidestar!

*Flip-mirror Finder* This is an essential accessory for CCD imaging. It enables the CCD camera field to be centred in an eyepiece mounted at right angles to the main optical axis. The mirror is then flipped, and the light is focused onto the CCD chip.

*Digital Setting Circles* I have recently fitted shaft encoders and a JMI MicroMAX computer to the telescope. Owing to the inexorable advance of long-sightedness, I find the LED readout on the computer to be much easier to read than the analogue setting circles on the mount, particularly when the latter need to be read in dim red light. The internal object database of the computer of these digital setting circles is small, but it is simple to look up the celestial coordinates of an object that is not in the database and then either use the circles in the conventional manner or use the "guide" mode of the computer.

*Power Supplies* The internal 9V battery is supposedly adequate for about 50 hours of driving the telescope at the sidereal rate. This diminishes greatly when the declination motor is used and/or when drive corrections are made using the hand paddle. Frequent replacement of alkaline batteries is expensive, so a rechargeable power supply is a desirable addition. I have two. One is a simple home-made device which consists of ten RX6 (AA) NiCd cells in series, mounted in a project box, which itself is attached to the wedge by Velcro strips. The ten NiCd cells give a nominal voltage of 12 V, and can be removed and recharged in a normal domestic battery charger. This power supply is adequate for several evenings of purely visual observing before it needs recharging. When I wish to power more than the telescope, for example when I am using a CCD camera and a laptop computer with the telescope, more power is needed and the home-made supply is inadequate. In these circumstances I use a Draper PP12V portable power pack, which contains a 12V 12Ah rechargeable lead–acid battery, sufficient for a full night's power needs. The power pack also has an integral lamp which, when covered with a red lens, is suitable for illuminating the telescope during set-up at the beginning of an observing session or during take-down at the end.

**Figure 10.5.** Field tripod with strengthening accessory tray and vibration suppression pads.

*Field Tripod* The standard telescope come with a table-top mount. The instrument becomes much more versatile if it is mounted on a tripod, and a lightweight, dedicated field tripod is available, upon which the table-top mount fixes directly. The tripod is made much more rigid if the optional accessory ray is fitted (see Fig. 10.5).

*Vibration-suppression Pads* These each consist of a rigid plastic "cup", into which fits another cup of sorbothane. The sorbothane cup contains a disc of rigid plastic. They fit between the tripod feet and the ground (see Fig. 10.5) and very effectively damp any vibration that might otherwise be transmitted to the tripod, be it by an accidental knock or a gust of wind. They are essential for photography or imaging at the Cassegrain focus.[3]

---

[3] This is often mistakenly called "prime focus" photography. The prime focus is the focus of the primary mirror, which is optically inaccessible in a telescope of this type unless the secondary mirror is removed. The "normal" focal plane of this telescope is properly called the Cassegrain focus or the Schmidt–Cassegrain focus.

*Hartmann Mask* This is a simple focusing aid, consisting of a full-aperture mask with two widely spaced holes in it. When the telescope is out of focus, a bright star will form two images. These will merge as focus is attained.

*35mm SLR Camera* The camera I use for astrophotography is a venerable Canon FTb. Its advantages for astrophotography include its excellent mechanical shutter and the mirror lock-up facility. For piggyback photography there is a large selection of good second-hand lenses available inexpensively. Although replacement focusing screens are not readily available, it is possible to attain good focus by focusing a bright star off the central microprism ring. Alternatively, a Hartmann mask permits excellent focus to be attained. With the reducer/corrector in place, a 35mm film frame ($24 \times 36$ mm) covers an area of sky of $1.7 \times 2.6°$.

*CCD Camera* I have recently acquired a Starlight Express MX5 CCD camera. This has a chip of $290 \times 500$ pixels in an area of $3.7 \times 4.9$ mm. Given an ideal 2 arc seconds per pixel for best resolution, a focal length of 1154 mm (45.4 inches) is optimum. With the focal reducer in place, the focal length of the C5 is about 800 mm (32 inches), giving just under 3 arc sec per pixel and a field of $15 \times 19.5$ arc minutes.

*Infrared Filter* Because CCDs are sensitive to infrared, an infrared blocking filter is an essential accessory for CCD imaging.

*Oxygen-III Filter* Among my favourite visual targets are planetary nebulae. The O-III filter enhances the visibility of these objects.

*Case* There are proprietary cases available for the C5, but these have always seemed to me to be overpriced. I have found that a ZAG Mobile Tool Chest, without modification, accommodates the drive base and optical tube assembly. All that is necessary is to cut some foam, preferably good-quality polyethylene foam, to fit. The integral tool tray, also fitted with foam, houses numerous accessories (see Fig. 10.6). Unfortunately, when digital setting circles are fitted, the cover for the declination encoder means that the instrument no longer fits neatly, and it is necessary to remove the central section of the accessory tray. The case's integral

**Figure 10.6.** The C5+, along with many accessories, fits into a toolbox.

wheels and pull-out handle enhance the mobility of the cased telescope, and the chest itself is remarkably sturdy, taking my 90 kg (200 lb) weight with ease, thus allowing it to double as a seat.

# Eyepieces

An advantage of an $f/10$ telescope is that the relatively acute light cone is not as demanding of eyepieces as are the more obtuse light cones of, for example, "fast" Newtonians. Even with the $f/6.3$ focal reducer in place it is possible to use moderately priced eyepieces. As I have more than one telescope, it is quite possible that, when I am observing together with other people, more than one eyepiece is in use at any given time. This is why I have not "rationalised" the apparent duplicate combinations. My current eyepiece selection consists of:

*32mm Plössl* With the focal reducer in place, this gives a magnification of ×25 with a 5mm exit pupil and a field of view of nearly 2°. This makes it an ideal choice for observing large, faint deep-sky objects. Used without the focal reducer, the entire disk of the Moon or the Sun (observed with the solar filter, of course!) is nicely framed in the field.

*25mm Kellner* This was irresistibly cheap at a sale, and is my least favourite eyepiece, not least because it suffers from excessive ghosting when used on bright objects. However, it is acceptable at $f/10$, and is my "public viewing" eyepiece (the purpose for which I bought it) since I do not become too upset when the eye lens picks up dollops of mascara.

*12.5mm Orthoscopic* I have had this one for a very long time and keep it because of its excellent optical quality. I use it on the Moon and planets when the seeing is too poor to permit a higher power than the ×100 it provides at $f/10$. It gives image of better contrast than does my 9mm Plössl, and is good for occultation timing and for glimpsing those smaller deep-sky objects of low surface brightness that benefit from higher magnification.

*9mm Plössl* This one is a reasonable example of its type. It has almost the same field of view as the 12.5mm orthoscopic, but is not quite as contrasty, despite the higher magnification. At $f/10$ there is no contest, but the 9mm is slightly better to the edge at $f/6.3$. This is my "high power public viewing" eyepiece, and I shall probably retain it for this purpose.

*6.3mm Plössl* On the face of it, this eyepiece (which came to me as a gift) is wasted since it duplicates the 12.5mm/Barlow combination. Like the orthoscopic, I keep it because it is of very good quality.

*9mm Illuminated Reticle Guiding Eyepiece* This eyepiece is of the Plössl design and has the advantage of a movable reticle. Two screws in the barrel move the reticle in two directions at right angles, thus making it easier to frame the star at the centre of the bull's-eye/double cross-hair markings. The illuminator is dimmable, making it easier to see fainter guide stars.

*Barlow Lens, ×2* This is an apochromatic design of excellent quality – anything less is unacceptable in a Barlow.

Given that I never use the Kellner for serious observing, the obvious gap in the above selection is something in the 20mm region. I am at present undecided between getting a 20mm Erfle (an eyepiece of which I have become enamoured) and trying one of the medium-power zoom eyepieces which are becoming more common and whose quality is reported to be very good. The Erfle–Barlow combination would effectively render the 9mm Plössl superfluous.

# Set-up and Use

The telescope is very transportable, and is therefore easy to take dark observing sites. I keep the tripod and wedge attached to each other unless I am intending to table-mount the telescope.

The most time-consuming aspect of setting up is polar alignment. The first task is to align the mount, before the telescope is placed upon it. A firm footing for the tripod is essential and, once the legs are spread and I have bolted the accessory tray firmly into place on the struts, I place the anti-vibration pads under the tripod feet. Next, I adjust the legs so as to centre the bubble in the bubble-level. The mount is now level, and the latitude scale on the equatorial wedge will be correct.

I have some added some white markings to the mount which aid in setting the azimuth, which can be adjusted a few degrees by loosening the Allen-head bolts that attach the wedge to the tripod, and rotating it about a vertical axis. I use a compass, with the bezel offset to the magnetic declination of my observing site, to align the markings on the mount to due north. I check the bubble-level, adjusting the legs if necessary.

The drive-base, to which I keep the telescope permanently attached, is secured to the wedge with three bolts. Once it is secure, I recheck the level and, if it is centred, the telescope is sufficiently well aligned for visual observing. The subsequent initialisation of the digital setting circles will then compensate for any misalignment when they are used to locate objects.

Apart from visual work, this compass-and-level alignment is not adequate for anything other than very short photographic exposures through the telescope or for short focal-length piggyback photography. A better alignment can be obtained by setting the declination to 90° (initially using the physical setting circle, latterly using marks on the declination bearing on the fork that I scribed for this purpose) and adjusting the mount so as to offset Polaris by an estimated 44 arc minutes in the appropriate direction from the centre of the field of view (i.e. away from the direction of β Ursae Minoris). With practice this can be estimated to within a few arc minutes using the focal reducer and the 32mm Plössl.

This improves the situation for photography, and is suitable unless I wish to take long-exposure photographs. For this purpose the mount must be aligned using the "star drift" method. This is extremely time-consuming, usually extending set-up time to well over an hour. The biggest source of frustration is the lack of facility on the mount for fine adjustment in altitude. This means that altitude adjustment can be accomplished only by adjusting the northerly tripod leg – not a very satisfactory arrangement! Despite some effort, I have yet to find a workable solution to this problem and would be delighted if one becomes available. The two advantages of the long time taken to align the mount are that it affords adequate time for the telescope optics to reach thermal equilibrium, and for the eyes to become dark-adapted. A quick check of collimation to confirm that it hasn't shifted is usually all that is required. It very rarely needs adjustment, and then only by a minuscule amount; this is testimony to the excellent mechanical construction of the optical tube assembly.

I then attach whatever accessories – camera, guidescope, counterweights, etc. – I intend to use, ensure that the telescope is balanced,[4] and the evening's session can finally begin (see Fig. 10.7)!

The notes I made after my first session with this telescope give an impression of what can be expected, and demonstrate why the C5 has such a good reputation:

> Just after sunset I went back outside to a ready-cooled telescope, put a 32mm Plössl (×39) in the diagonal, and found Jupiter – the telescope snapped into focus and both equatorial belts were clearly visible. I tried the 6.3mm Plössl (×198) and again achieved a well-defined focus. The equatorial belts showed ragged edges, that of the southern edge of the SEB being particularly clear. Temperate belts (I haven't a clue which – I'm not a planetary observer) were also visible. Diffraction rings showed around the Galilean moons.
>
> Saturn was similarly impressive at ×198. The Cassini Division was immediately clearly visible, and I was surprised to see the planet's shadow on the E side of

---

[4] "Balancing" a telescope, when it is to be driven, actually means having it very slightly out of balance about the RA axis in order that the RA drive is working "uphill". This results in better tracking.

**Figure 10.7.** The fully
equipped telescope.

the ring – its width must be close to the limit of resolution of the telescope.[5] Markings were visible on the
planet's surface, with the polar regions distinctly
darker than the equatorial region.

The diffraction rings around the Jovian moons suggested that a star-test was in order. Aldebaran was
conveniently placed and, at ×198, showed a bright Airy
disc surrounded by four diffraction rings, the inner-
most of which was quite bright. Defocused images
revealed some undercorrection. At this point it may be
pertinent to note that focusing is extremely smooth
and that I could not detect any image-shift – I'll try it
another time with a reticle in a high-power eyepiece.

At this point I was called indoors. When I returned
some 20 minutes later, I was pleasantly surprised to
find that Aldebaran had hardly shifted in the field of
view, suggesting that the tracking was good and
that my daylight polar aligning was (fortuitously?)
accurate.

---

[5] A later check showed that the planet's shadow had a
maximum width of 0.9 arc second, which is indeed very close
to the theoretical resolution of a 125mm aperture.

After an evening meal, I returned to the telescope under darker but light-polluted skies. M42 was emerging from behind the branches of my neighbour's birch tree, and at ×39 showed some pleasing structure. The four stars of the Trapezium were clearly resolved. I tried an O-III filter and almost gasped at how much more structure I could see.

Chasing objects nearer the zenith, particularly between the zenith and the pole, proved to be an interesting task and reminded me why I like German equatorial mounts – the fork tine and base always seemed to be where I wanted to put my face. A red-dot finder would be a boon here, allowing the head to be farther from the telescope than the $6 \times 30$ finder.

At ×39, M31 was about what I expected with a 3.2mm exit pupil, suggesting that the $f/6.3$ focal reducer might be an early purchase – it would give a 5mm exit pupil with the 32mm Plössl. The open clusters in Auriga were as beautiful as ever, and a cruise through Perseus and Cassiopeia revealed several more small clusters which I enjoyed but didn't bother to identify.

By this time the Moon was rising over the neighbouring bungalows and the sky was brightening significantly. Some dew was beginning to form (must make a dewcap, or this corrector plate is going to be almost permanently wet!), and it seemed a good time to call a break.

The British weather precluded further observations for over a week, but when the cloud next cleared a very spotty Sun was revealed through the solar filter. At $f/10$, the Sun is nicely framed in the field of the 32mm Plössl and also on the frame of the 35mm camera. Photographs again revealed the very good optics of this telescope, and also the quality of the Baader Astro-Solar film (see Fig. 10.8).

The next nocturnal outing for the telescope was a practice run for a Messier marathon attempt with the Wessex Astronomical Society at Worth Hill Observatory. This gave me the opportunity to directly compare it with other telescopes on the same objects in precisely the same conditions. The C5, now fitted with the reflex finder and the reducer/corrector, again gave a good account of itself, and the quality of its optics drew favourable comments from several experienced observers. Another 125mm (5-inch) catadioptric at the same site was, despite its computerised pointing, unable to reveal M101 to its owner. This object was difficult, but not greatly so, with the 32m Plössl in the C5.

**Figure 10.8.** Sunspots photographed through the solar filter; ISO 400 Kodak EliteChrome film, exposure 1/500 second.

Piggyback photography, using the telescope as a guidescope, produces some fine results as long as polar alignment of the fork mount is sufficiently accurate (see Fig. 10.9). The drive errors are such that they are simple to "guide out" with the hand paddle. Long-exposure prime-focus photography (strictly speaking, pho-

**Figure 10.9.** The region of Orion's Belt and M42 ; ISO 400 film, 200mm f/4 lens, exposure 6 minutes.

**Figure 10.10.** The Double Cluster in Perseus; ISO 400 film, Cassegrain focus, exposure 20 minutes.

tography at the Cassegrain focus) places more demands on the mount than does piggyback photography. It is essential to use the anti-vibration pads and to use the accessory tray to improve the rigidity of the tripod. Guiding is essential. The radial guider is compact and puts less strain on the mount than does the guidescope, but the guidescope is far easier to use. Figure 10.10 shows an example of what it is possible to achieve.

More recently I have moved into CCD imaging. I find that the Starlight Express MX5 works well with the C5+ and, surprisingly, that about a third of all unguided exposures of 60 seconds are of acceptable quality (see Fig. 10.11). However, the MX5 projects a long way from the OTA, potentially fouling the drive base, thus rendering unavailable a large region of the sky in the vicinity of the celestial pole.

M13 2000 Jun 09 02:04 UT 30s

**Figure 10.11.** M13 in Hercules; Starlight Express MX5 CCD camera, C5 at f/6.8, exposure 30 seconds.

# Conclusion

The C5 is a very good OTA and makes the basis of a useful and productive portable set-up. The C5+ drive base is very good and is more than adequate for visual work. The single-tine fork imposes a severe limitation on the accessories that can be carried on the telescope and must be counterweighted when cameras or CCDs are used. The combined weight of counterweights and accessories then pushes the carrying capacity of the mount to near its limit. As a small imaging platform, a good-quality German equatorial mount, such as the Vixen Great Polaris, would provide a much more stable base and would not suffer from the fork-mount's limitations in the region of the celestial pole. Given that the C5 OTA is available separately and that the C5+ is becoming increasingly difficult to obtain, mating it with a good German equatorial mount is probably the best option if you wish to consider photography or CCD imaging.

# Part IV

# Radio

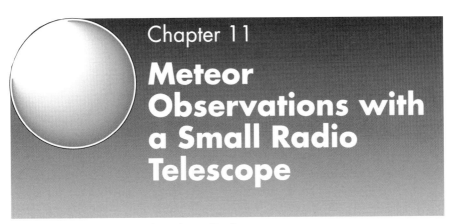

# Chapter 11

# Meteor Observations with a Small Radio Telescope

Stephen Tonkin

It is probably my imagination, but for years it has seemed to me that most of the really active meteor showers reach their maximum when the Moon is unfavourable, the Sun is above the horizon, or the sky is overcast. But since I began to observe these showers with a small radio telescope, these unfavourable conditions have mattered little to me.

As a meteor passes through Earth's atmosphere, some of its kinetic energy is transferred to the surrounding gases, causing their atoms to ionise. Because this train is ionised, it is able to reflect radio signals. VHF radio signals are "line of sight", so distant transmitters are normally undetectable if they are below the observer's horizon. However, some of the scatter of the radio signal off the meteor's ionisation train will reach the observer and, if it is sufficiently strong, will be detectable by a radio receiver (see Fig. 11.1). This is perhaps the simplest form of radio astronomy, requiring only very simple equipment and largely unaffected by the radio noise that plagues radio observations of sources outside the Solar System. Radio detection of meteors has an advantage over visual detection in that many more meteors are detectable; just how many more depends upon the sensitivity of the receiver, but even my very modest set-up detects about five times as many as a visual count. This is poor – a better antenna would at least double this.

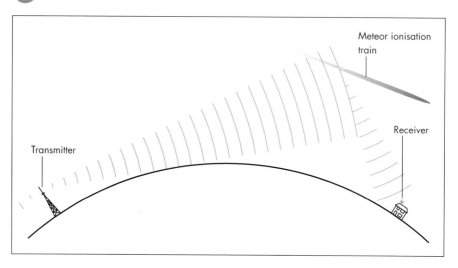

**Figure 11.1.** Forward scatter of radio signals by a meteor's ionisation trail.

My first attempts at meteor detection involved nothing more than a domestic FM radio receiver. Having woken one morning in mid-December, I retuned the bedside radio to a BBC Radio 4 transmitter that is below my horizon and proceeded to listen, with mounting excitement, to the random momentary snatches of the *Today* programme that were reflected off the ionisation trails of Geminid meteors. After expressing a dutiful interest in a longer-lasting ionisation trail, my wife's demeanour suggested that she was somewhat less impressed than I was, and shortly afterwards the radio was retuned to a local transmitter.

As it was only three weeks from a predicted midday maximum of the 1998 Quadrantid shower (3 January), I decided to concoct a cheap, small and simple detection system. The heart of this is a Yupiteru MVT-7100 hand-held multi-band scanning radio (see Fig. 11.2). This little receiver has a nominal frequency range of 100 kHz to 1650 MHz (but this is unreliable outside the range 530 kHz to 1300 MHz) and, what was important for the use to which I was intending to put it, it has receiving modes that include single side band (SSB). A number of European and Irish television stations (all below my horizon) transmit an SSB VHF signal with a carrier frequency in the old UK TV1 band; the frequencies are 48.2502, 53.757, 53.7396, 53.7604 and 55.2604 MHz.

In order to receive a signal, the receiver must be connected to an antenna. The Yupiteru MVT-7100 is supplied with a telescopic whip antenna as standard. While this type of antenna is useful, not least because its length is adjustable to match it to the frequency of the

**Figure 11.2.** The Yupiteru MVT-7100 receiver.

signal being received (a whip is a quarter-wave element), it is also extremely susceptible to damage – in my experience, this is the fate of most telescopic whips. The helical ("rubber duck") broadband antenna that is now very popular for hand-held receivers is more robust, but does not perform as well as a whip. Since the receiver does not have be moved *while* the meteor count is taking place, the antenna used need not be what is normally considered to be mobile.

For simplicity I chose to make a simple dipole antenna optimised for approximately 50 MHz. A dipole consists of two quarter-wave elements. At 50 MHz these would each be about 1.5 m (5 ft) long, but they need to be shortened to 1.4 m in order to match the antenna's impedance to that of the receiver, in this case 50 Ω. At nearly 3 m (10 ft) long, the resulting antenna cannot be considered to be conveniently mobile, but it is something for which space can be found in an attic, or even along the top of a wall. I made mine of some insulated copper wire that I happened to have available. It resides in the attic and connects to the receiver with a length of coaxial cable, but I can easily remove it, roll it up, and take it to other locations.

I had been told of, but had never actually heard for myself, the ping of a meteor detection. I did not make my dipole antenna until the morning of the Quadrantid peak. Sitting in the attic, I tuned the receiver to 53.757 MHz, turned the squelch off, and listened apprehensively to the background radio noise. The first ping, which sounded exactly as it had been described to me, was a very exciting moment. The characteristic sound of a meteor ping is a sudden high-pitched note that is sustained for a short period (depending upon the duration of the meteor's ionisation train) before falling exponentially in amplitude and simultaneously also falling slightly in pitch (see Fig 11.3). If the rise in amplitude is not sudden, or if the ping is continuous, it is not a meteor but some other atmospheric effect that is the cause. During that first one-hour session I counted 149 pings of varying amplitude and duration, some lasting several seconds before dying away.

I am primarily a visual observer, and I used the small radio telescope sporadically over the following months. Its first public outing came on 12 August when a "family fun day" in a local park, at which my local astronomical society was represented, coincided with the broad peak of the Perseid shower. With the antenna held up on a jury-rigged support on the roof-rack of my car, and the

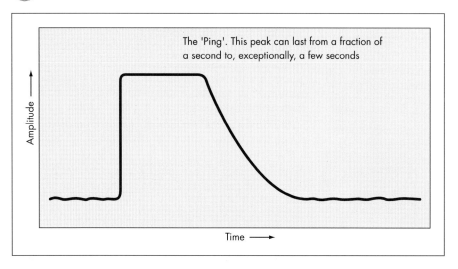

The 'Ping'. This peak can last from a fraction of a second to, exceptionally, a few seconds

Amplitude ⟶

Time ⟶

**Figure 11.3.**
A meteor ping.

eyes of the public protected by table-tennis balls at its ends, the pings began, not very strongly or frequently, but sufficient to generate excitement among those astronomers present who had never before heard the phenomenon. I think it is fair to say that most members of the public were more amused (and bemused) by our excitement and enthusiasm than were educated by the pings – people usually want to *see* things.

That night our astronomical society held a public meteor watch at Worth Hill Observatory on the Isle of Purbeck. An illustrated talk as twilight fell prepared those members of the public who were present. I rigged my antenna by stringing it out along a hedgerow, and began to listen to the pings. What did excite members of the public was the correlation of loud pings with visible meteors.

There are two obvious areas for improvement: antennas and recording, both of which are on my "to do" list. First, the dipole antenna is no more than adequate to demonstrate the principle. A high-gain antenna would be the radio equivalent of increasing light-gathering power (but *not* resolution!) by increasing the aperture of an optical telescope. A steerable Yagi antenna, mounted on a mast on the roof of the house, would have much higher gain and would record many more meteors. Different antennae could be used for different purposes, such as detecting radio signals from the Sun (e.g. solar flare detection), Jupiter, or some of the more powerful radio sources outside the Solar System (see Table 11.1). Because of its size, a

**Table 11.1.** Antenna sizes for various purposes

| Object | Frequency (MHz) | $\frac{1}{4}$-wave (m) | 50 Ω dipole length (m) |
|---|---|---|---|
| Jupiter | 20 | 3.72 | 7.08 |
| Meteor (TV1) | 50 | 1.49 | 2.84 |
| Meteor (FM radio) | 100 | 0.745 | 1.42 |
| Sun, Galaxy | 150 | 0.496 | 0.942 |

Jupiter antenna could not be conveniently and safely mounted on the roof of an ordinary home.

Second, connecting a tape recorder to the receiver is a trivial exercise, but does no more than permit me to do a manual count at a later date. Continuous monitoring and counting is something that can be done by connecting the receiver to a computer via a suitable analogue to digital converter (ADC). An ADC circuit can either be built onto a PC expansion card or, more simply, as a "box" that connects the receiver to one of the computer's ports. Once this is done, public domain software can be obtained to perform the counting and recording. Depending upon the receiver you use, you may need to disable the automatic gain control – consult the manufacturer before doing this as you may invalidate any warranty.

A small, portable, multipurpose radio telescope offers a great deal of potential in limited areas of radio astronomy, particularly those where change is detectable, such as meteor or solar flare observations. In other areas, such as detection of galactic radio sources, its use is essentially limited to demonstrating that they exist.

# The Contributors

**Kevin Daly** is a thirty-seven-year old executive and a married father of four, currently residing in Wolcott, CT. He recently began in astronomy using 7 × 50 binoculars and currently uses these plus a Celestron 114mm (4.5-inch) Newtonian reflector, a Criterion 200mm (8-inch) *f*/7.5 Newtonian reflector (both on German equatorial mounts), a Celestron 60mm (2.4-inch) refractor, and an Intes 100mm (4-inch) Maksutov–Cassegrain spotting scope (his "travel scope"). He is an active member of the Mattatuck Astronomical Society in Waterbury, CT. He also hosts a very successful Web page (http://hometown.aol.com/kdaly10475/index. html) dedicated to helping newcomers to astronomy.

**Dwight Elvey,** born in 1947, is an electronics engineer whose hobbies include astronomy, telescope-making, collecting old computers, sailing, and model slope gliders. He is a member of the Santa Cruz Astronomy Club and the San Jose Astronomy Association. He took up astronomy in the mid-1990s and most enjoys looking at distant galaxies and nebulas. His interests in the science of telescopes, the mechanics of telescope optics and the construction of telescopes are on par with his love of observing.

**Jay Reynolds Freeman** is an amateur astronomer who loves the deep sky. He has logged over 11,000 observations of almost 5,000 objects, and used some thirty telescopes and binoculars enough to know them well. His life-long astronomical interests led to a PhD in physics, studying the interstellar medium with instruments on an Apollo spacecraft, but his thesis work used only extreme ultraviolet light, so he retains amateur status in visual wavelengths. He is active on the Internet newsgroup, "sci.astro.amateur", and has published a few articles in *Sky & Telescope* magazine. He works as a computer programmer, lives in Palo Alto, California, and doesn't know what he wants to be if he grows up.

**Robert Hatch** (rob@gbcgq.freeserve.co.uk) was born in 1951 in North London but has lived in Dorset for most of his life. He has a degree in chemistry and physics from London University. An interest in astronomy developed in the 1960s, and in 1969 he became a member of the Wessex Astronomical Society (then known as the Bournemouth,

Poole and District Astronomical Association), for which he has served a term as Chairman. His interests include listening to classical music, technical reading, computing, radio and electronics, (he holds an amateur radio licence), and works in Poole, Dorset, as part of a quality-assurance team in the electronic and manufacturing field. He is active in his local United Reformed Church, and holds definite views on life elsewhere in the Universe (as well as after death!), but (just for fun) is contributing to the SETI@home project.

**Dave Mitsky's** (djm28@psu.edu) interest in astronomy began as a young child and eventually resulted in the purchase of a small Newtonian reflector from Edmund Scientific, a telescope that served him well for many years. At one time he considered majoring in astronomy but finished college with a degree in biophysics instead. He considers himself to be primarily an aficionado of the deep sky, but his interests run the gamut of amateur astronomy and include meteor, lunar, planetary and solar observing. He currently owns five telescopes, holds a number of Astronomical League Observing Club awards, and is an active member of two Pennsylvania astronomy clubs – the Astronomical Society of Harrisburg (ASH), where he has served as vice-president , and the Delaware Valley Amateur Astronomers (DVAA). One of his achievements as an amateur astronomer is the 1997 discovery of the Walter Sunset Lunar Ray using the 430mm (17-inch) $f/15$ classical Cassegrain at the ASH Naylor Observatory.

**Stephen Tonkin** (http://www.aegis1.demon.co.uk) grew up under the dark skies of tropical Africa and has been interested in astronomy for as long as he can remember. He is a keen telescope maker and also takes delight in modifying commercial offerings. By profession he is a teacher, and his hobbies include story-telling, singing and making music, and amateur dramatics. He is the current chairman of the Wessex Astronomical Society.

**Tim Tonkin** grew up with a father (Stephen) obsessed with astronomy, which has therefore always been an influence on his life. By the age of eight he had made an independent binocular discovery of a Messier object and, two years later, he made his first reflecting telescope. When he leaves school, he intends to go to university and study law. He is a keen cricketer.

**Mike Weasner** is the Webmaster of Weasner's Mighty ETX Site (http://www.weasner.com/etx). He grew up in Seymour, Indiana, USA. A brother started him in astronomy when he was six years old. Mike earned an astrophysics degree from Indiana University, Bloomington, Indiana; did some postgraduate work in meteorology at the University of Wisconsin, Madison; and served in the United States Air Force as a fighter pilot, an instructor, and a manager in the

Air Force's Space Shuttle Program Office. He now works at an aerospace company. His hobbies include astronomy, science fiction, and computers. He enjoys classical music, science fiction movie soundtracks, and old radio shows such as *X Minus One* and *The Lone Ranger*.

# Choosing and Using a Schmidt-Cassegrain Telescope

## A Guide to Commercial SCTs and Maksutovs

*Rod Mollise*

The commercial availability of high-quality Schmidt-Cassegrain Telescopes (SCTs) has been a major factor in the upsurge of interest in amateur astronomy.

Modern SCTs and other CATs - catadioptric (lens-mirror) telescopes - are incredibly capable tools for amateur astronomers.  Models range from 3½-inch portable instruments to 16-inch giants, and all sizes now have computer-control at least as an option.  But because they are real scientific instruments, using them can sometimes be tricky. Rewarding but tricky.

Choosing and using a Schmidt-Cassegrain telescope is exactly what this book is about.  It's a compilation of the hints, tips and general wisdom of many experienced SCT users, distilled by Rod Mollise, who has used SCTs himself for some 25 years.

**190 pages**
**Softcover**
**ISBN 1-85233-631-5**